Pocket Book of Integrals and Mathematical Formulas
2nd Edition

Ronald J. Tallarida

CRC Press
Boca Raton Ann Arbor London Tokyo

Library of Congress
Cataloging-in-Publication Data

Tallarida, Ronald J.
Pocket book of integrals and mathematical formulas / Ronald J. Tallarida.—2nd ed.
p. cm.

Includes bibliographical references and index.
ISBN 0-8493-0142-4
1. Integrals—Tables. 2. Mathematics—Tables. I. Title.
QA310.T35 1992 92-14684
510′.212—dc20 CIP

Direct all inquiries to CRC Press, Inc., 2000 Corporate Blvd. N.W., Boca Raton, Florida 33431.

2nd Edition

International Standard Book Number 0-8493-0142-4

Library of Congress Card Number 92-14684
Printed in the United States of America 3 4 5 6 7 8 9 0
Printed on acid-free paper

Preface to the Second Edition

This second edition has been enlarged by the addition of several new topics while preserving its convenient pocket size. New in this edition are the following topics: z-transforms, orthogonal polynomials, Bessel functions, probability and Bayes' rule, a summary of the most common probability distributions (Binomial, Poisson, normal, t, Chi square and F), the error function, and several topics in multivariable calculus that include surface area and volume, the ideal gas laws, and a table of centroids of common plane shapes. A list of physical constants has also been added to this edition.

I am grateful for many valuable suggestions from users of the first edition, especially Lt. Col. W. E. Skeith and his colleagues at the U.S. Air Force Academy.

R.J.T.
Philadelphia, 1992

Preface to the First Edition

The material of this book has been compiled so that it may serve the needs of students and teachers as well as professional workers who use mathematics. The contents and size make it especially convenient and portable. The widespread availability and low price of scientific calculators have greatly reduced the need for many numerical tables (e.g., logarithms, trigonometric functions, powers, etc.) that make most handbooks bulky. However, most calculators do not give integrals, derivatives, series, and other mathematical formulas and figures that are often needed. Accordingly, this book contains that information in addition to a comprehensive table of integrals. A section on statistics and the accompanying tables, also not readily provided by calculators, have also been included.

The size of the book is comparable to that of many calculators and it is really very much a companion to the calculator and the computer as a source of information for writing one's own programs. To facilitate such use, the author and the publisher have worked together to make the format attractive and clear. Yet, an important requirement in a book of this kind is accuracy. Toward that end we have checked each item against at least two independent sources.

Students and professionals alike will find this book a valuable supplement to standard textbooks, a source for review, and a handy reference for many years.

Ronald J. Tallarida
Philadelphia

About the Author

Ronald J. Tallarida holds B.S. and M.S. degrees in physics/mathematics and a Ph.D. in pharmacology. His primary appointment is as Professor of Pharmacology at Temple University School of Medicine, Philadelphia; he also serves as Adjunct Professor of Biomedical Engineering (Mathematics) at Drexel University in Philadelphia.

He received the Lindback Award for Distinguished Teaching in 1964 while in the Drexel mathematics department. As an author and researcher, Dr. Tallarida has published over 150 works, including 7 books. He is currently the series editor for the Springer-Verlag Series in Pharmacologic Science.

Greek Letters

α	A	Alpha
β	B	Beta
γ	Γ	Gamma
δ	Δ	Delta
ϵ	E	Epsilon
ζ	Z	Zeta
η	H	Eta
θ	Θ	Theta
ι	I	Iota
κ	K	Kappa
λ	Λ	Lambda
μ	M	Mu
ν	N	Nu
ξ	Ξ	Xi
o	O	Omicron
π	Π	Pi
ρ	P	Rho
σ	Σ	Sigma
τ	T	Tau
υ	Υ	Upsilon
ϕ	Φ	Phi
χ	X	Chi
ψ	Ψ	Psi
ω	Ω	Omega

The Numbers π and e

π	=	3.14159	26535	89793
e	=	2.71828	18284	59045
$\log_{10}e$	=	0.43429	44819	03252
$\log_{e}10$	=	2.30258	50929	94046

Prime Numbers

2	3	5	7	11	13	17	19	23	29
31	37	41	43	47	53	59	61	67	71
73	79	83	89	97	101	103	107	109	113
127	131	137	139	149	151	157	163	167	173
179	181	191	193	197	199	211	223	227	229
233	239	241	251	257	263	269	271	277	281
...				...					...

Important Numbers in Science (Physical Constants)

Avogadro constant (N_A)	6.02×10^{26} kmole^{-1}
Boltzmann constant (k)	1.38×10^{-23} J°K^{-1}
Electron charge (e)	1.602×10^{-19} C
Electron, charge/mass, (e/m_e)	1.760×10^{11} C·kg^{-1}
Electron rest mass (m_e)	9.11×10^{-31} kg (0.511 MeV)
Faraday constant (F)	9.65×10^{4} C·mole^{-1}
Gas constant (R)	8.31×10^{3} J·°K^{-1}·kmole^{-1}
Gas (ideal) normal volume (V_o)	22.4 m^3·kmole^{-1}
Gravitational constant (G)	6.67×10^{-11} N·m^2·kg^{-2}
Hydrogen atom (rest mass) (m_H)	1.673×10^{-27} kg (938.8 MeV)

Neutron (rest mass) (m_n)	1.675×10^{-27} kg (939.6 MeV)
Planck constant (h)	6.63×10^{-34} J·s
Proton (rest mass) (m_p)	1.673×10^{-27} kg (938.3 MeV)
Speed of light (c)	3.00×10^{8} m·s^{-1}

Contents

3 Trigonometry

4 Analytic Geometry

5 Series

6 Differential Calculus

7 Integral Calculus

1 Elementary Algebra and Geometry

Algebra

1. Fundamental Properties (Real Numbers)

$a+b=b+a$	Commutative Law for Addition
$(a+b)+c=a+(b+c)$	Associative Law for Addition
$a+0=0+a$	Identity Law for Addition
$a+(-a)=(-a)+a=0$	Inverse Law for Addition
$a(bc)=(ab)c$	Associative Law for Multiplication
$a\left(\frac{1}{a}\right)=\left(\frac{1}{a}\right)a=1,\ a\neq 0$	Inverse Law for Multiplication
$(a)(1)=(1)(a)=a$	Identity Law for Multiplication
$ab=ba$	Commutative Law for Multiplication
$a(b+c)=ab+ac$	Distributive Law

DIVISION BY ZERO IS NOT DEFINED

2. *Exponents*

For integers m and n

$$a^n a^m = a^{n+m}$$

$$a^n / a^m = a^{n-m}$$

$$(a^n)^m = a^{nm}$$

$$(ab)^m = a^m b^m$$

$$(a/b)^m = a^m / b^m$$

3. *Fractional Exponents*

$$a^{p/q} = (a^{1/q})^p$$

where $a^{1/q}$ is the positive qth root of a if $a > 0$ and the negative qth root of a if a is negative and q is odd. Accordingly, the five rules of exponents given above (for integers) are also valid if m and n are fractions, provided a and b are positive.

4. *Irrational Exponents*

If an exponent is irrational, e.g., $\sqrt{2}$, the quantity, such as $a^{\sqrt{2}}$ is the limit of the sequence, $a^{1.4}, a^{1.41}, a^{1.414}, \ldots$.

- *Operations with Zero*

$$0^m = 0;\ a^0 = 1$$

5. *Logarithms*

If x, y, and b are positive and $b \neq 1$

$$\log_b(xy) = \log_b x + \log_b y$$

$$\log_b(x/y) = \log_b x - \log_b y$$

$$\log_b x^p = p \log_b x$$

$$\log_b(1/x) = -\log_b x$$

$$\log_b b = 1$$

$$\log_b 1 = 0 \qquad \textit{Note}: b^{\log_b x} = x.$$

- *Change of Base* $(a \neq 1)$

$$\log_b x = \log_a x \log_b a$$

6. *Factorials*

The factorial of a positive integer n is the product of all the positive integers less than or equal to the integer n and is denoted $n!$. Thus,

$$n! = 1 \cdot 2 \cdot 3 \cdot \ldots \cdot n.$$

Factorial 0 is defined: $0! = 1$.

- *Stirling's Approximation*

$$\lim_{n \to \infty} (n/e)^n \sqrt{2\pi n} = n!$$

(See also 9.2.)

7. Binomial Theorem

For positive integer n

$$(x+y)^n = x^n + nx^{n-1}y + \frac{n(n-1)}{2!}x^{n-2}y^2$$

$$+\frac{n(n-1)(n-2)}{3!}x^{n-3}y^3 + \cdots$$

$$+nxy^{n-1} + y^n.$$

8. Factors and Expansion

$$\begin{aligned}
(a+b)^2 &= a^2+2ab+b^2 \\
(a-b)^2 &= a^2-2ab+b^2 \\
(a+b)^3 &= a^3+3a^2b+3ab^2+b^3 \\
(a-b)^3 &= a^3-3a^2b+3ab^2-b^3 \\
(a^2-b^2) &= (a-b)(a+b) \\
(a^3-b^3) &= (a-b)(a^2+ab+b^2) \\
(a^3+b^3) &= (a+b)(a^2-ab+b^2)
\end{aligned}$$

9. Progression

An *arithmetic progression* is a sequence in which the difference between any term and the preceding term is a constant (d):

$$a, a+d, a+2d, \ldots, a+(n-1)d.$$

If the last term is denoted l $[=a+(n-1)d]$, then the sum is

$$s=\frac{n}{2}(a+l).$$

A *geometric progression* is a sequence in which the ratio of any term to the preceding term is a constant r. Thus, for n terms

$$a, ar, ar^2, \ldots, ar^{n-1}$$

The sum is

$$S=\frac{a-ar^n}{1-r}$$

10. *Complex Numbers*

A complex number is an ordered pair of real numbers (a,b).

Equality: $(a,b)=(c,d)$ if and only if $a=c$ and $b=d$
Addition: $(a,b)+(c,d)=(a+c,b+d)$
Multiplication: $(a,b)(c,d)=(ac-bd,ad+bc)$

The first element (a,b) is called the *real* part; the second the *imaginary* part. An alternate notation for (a,b) is $a+bi$, where $i^2=(-1,0)$, and $i=(0,1)$ or $0+1i$ is written for this complex number as a convenience. With this understanding, i behaves as a number, i.e., $(2-3i)(4+i)=8-12i+2i-3i^2=11-10i$. The conjugate of $a+bi$ is $a-bi$ and the product of a complex number and its conjugate is a^2+b^2. Thus, *quotients* are computed by multiplying numerator and denominator by the conjugate of the denominator, as illustrated below:

$$\frac{2+3i}{4+2i}=\frac{(4-2i)(2+3i)}{(4-2i)(4+2i)}=\frac{14+8i}{20}=\frac{7+4i}{10}$$

11. Polar Form

The complex number $x+iy$ may be represented by a plane vector with components x and y

$$x+iy=r(\cos\theta+i\sin\theta)$$

(see Figure 1.1). Then, given two complex numbers $z_1=r_1(\cos\theta_1+i\sin\theta_1)$ and $z_2=r_2(\cos\theta_2+i\sin\theta_2)$, the product and quotient are

product: $z_1z_2=r_1r_2[\cos(\theta_1+\theta_2)+i\sin(\theta_1+\theta_2)]$

quotient: $z_1/z_2=(r_1/r_2)[\cos(\theta_1-\theta_2)+i\sin(\theta_1-\theta_2)]$

powers: $z^n=[r(\cos\theta+i\sin\theta)]^n=r^n[\cos n\theta+i\sin n\theta]$

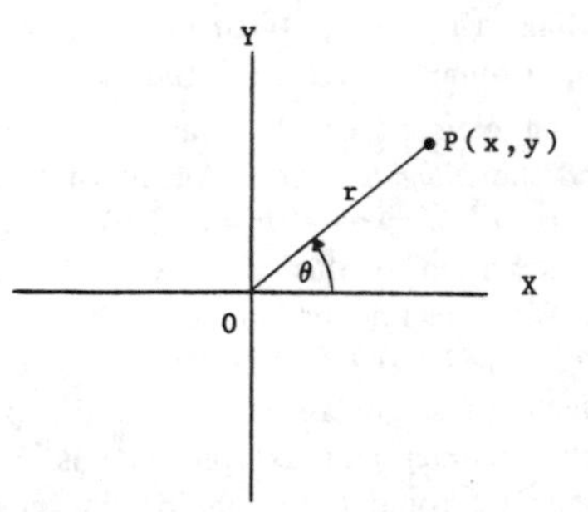

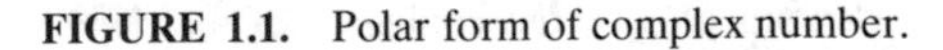
FIGURE 1.1. Polar form of complex number.

roots: $z^{1/n} = [r(\cos\theta + i\sin\theta)]^{1/n}$

$$= r^{1/n}\left[\cos\frac{\theta + k.360}{n} + i\sin\frac{\theta + k.360}{n}\right],$$

$$k = 0, 1, 2, \ldots, n-1$$

12. Permutations

A permutation is an ordered arrangement (sequence) of all or part of a set of objects. The number of permutations of n objects taken r at a time is

$$p(n,r) = n(n-1)(n-2)\ldots(n-r+1)$$

$$= \frac{n!}{(n-r)!}$$

A permutation of positive integers is "even" or "odd" if the total number of inversions is an even integer or an odd integer, respectively. Inversions are counted relative to each integer j in the permutation by counting the number of integers that follow j and are less than j. These are summed to give the total number of inversions. For example, the permutation 4132 has four inversions: three relative to 4 and one relative to 3. This permutation is therefore even.

13. Combinations

A combination is a selection of one or more objects from among a set of objects regardless of order. The number of combinations of n different objects taken r at a time is

$$C(n,r)=\frac{P(n,r)}{r!}=\frac{n!}{r!(n-r)!}$$

14. *Algebraic Equations*

- *Quadratic*

If $ax^2+bx+c=0$, and $a\neq 0$, then roots are

$$x=\frac{-b\pm\sqrt{b^2-4ac}}{2a}$$

- *Cubic*

To solve $x^3+bx^2+cx+d=0$, let $x=y-b/3$. Then the *reduced cubic* is obtained:

$$y^3+py+q=0$$

where $p=c-(1/3)b^2$ and $q=d-(1/3)bc+(2/27)b^3$. Solutions of the original cubic are then in terms of the reduced cubic roots y_1, y_2, y_3:

$$x_1=y_1-(1/3)b \qquad x_2=y_2-(1/3)b$$

$$x_3=y_3-(1/3)b$$

The three roots of the reduced cubic are

$$y_1=(A)^{1/3}+(B)^{1/3}$$

$$y_2=W(A)^{1/3}+W^2(B)^{1/3}$$

$$y_3 = W^2(A)^{1/3} + W(B)^{1/3}$$

where

$$A = -\frac{1}{2}q + \sqrt{(1/27)p^3 + \frac{1}{4}q^2},$$

$$B = -\frac{1}{2}q - \sqrt{(1/27)p^3 + \frac{1}{4}q^2},$$

$$W = \frac{-1 + i\sqrt{3}}{2}, \quad W^2 = \frac{-1 - i\sqrt{3}}{2}.$$

When $(1/27)p^3 + (1/4)q^2$ is negative, A is complex; in this case A should be expressed in trigonometric form: $A = r(\cos\theta + i\sin\theta)$ where θ is a first or second quadrant angle, as q is negative or positive. The three roots of the reduced cubic are

$$y_1 = 2(r)^{1/3}\cos(\theta/3)$$

$$y_2 = 2(r)^{1/3}\cos\left(\frac{\theta}{3} + 120°\right)$$

$$y_3 = 2(r)^{1/3}\cos\left(\frac{\theta}{3} + 240°\right)$$

15. *Geometry*

The following is a collection of common geometric figures. Area (A), volume (V), and other measurable features are indicated.

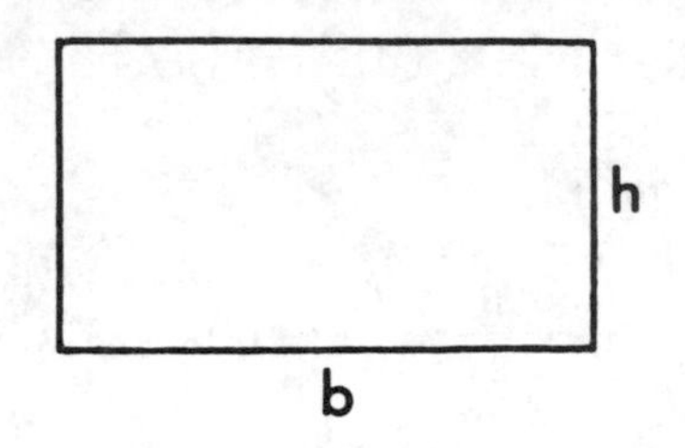

FIGURE 1.2. Rectangle. $A = bh$.

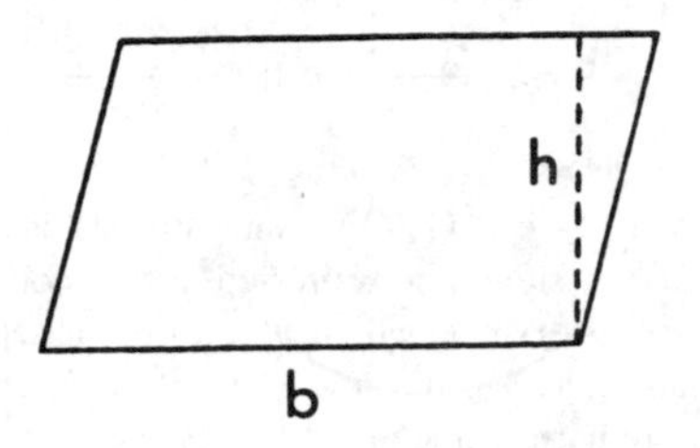

FIGURE 1.3. Parallelogram. $A = bh$.

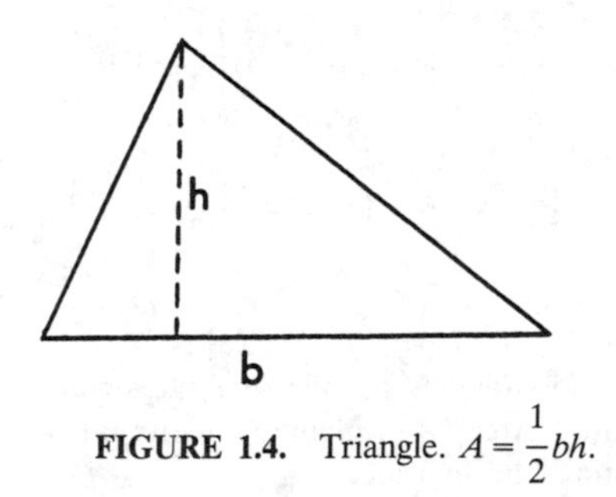

FIGURE 1.4. Triangle. $A = \frac{1}{2}bh$.

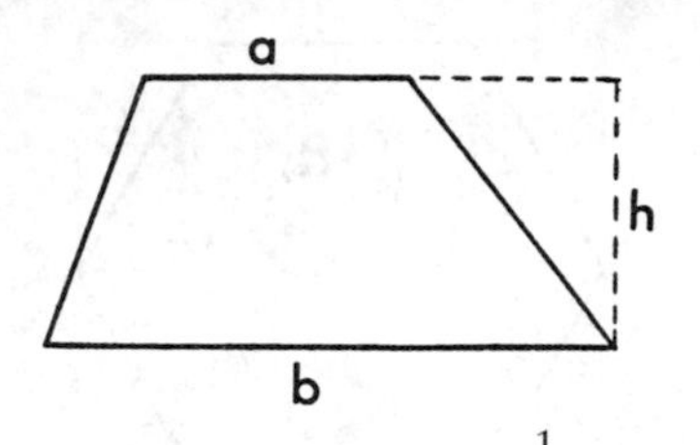

FIGURE 1.5. Trapezoid. $A = \frac{1}{2}(a+b)h$.

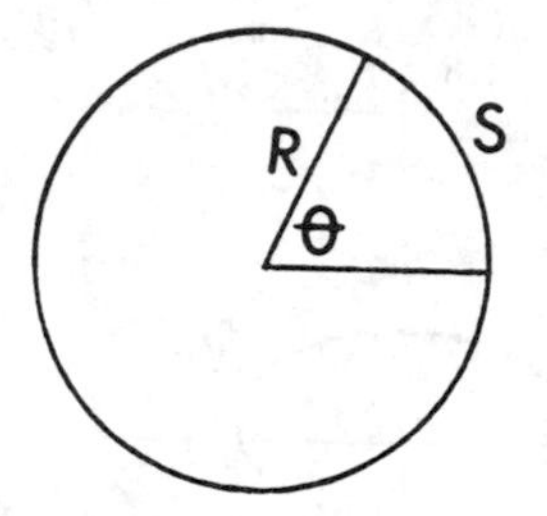

FIGURE 1.6. Circle. $A = \pi R^2$; circumference $= 2\pi R$; arc length $S = R\theta$ (θ in radians).

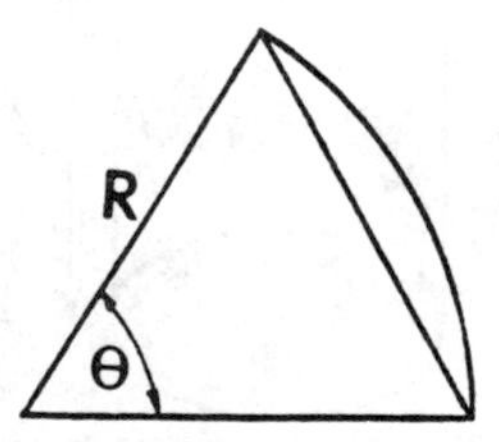

FIGURE 1.7. Sector of circle. $A_{\text{sector}} = \frac{1}{2}R^2\theta$; $A_{\text{segment}} = \frac{1}{2}R^2(\theta - \sin\theta)$.

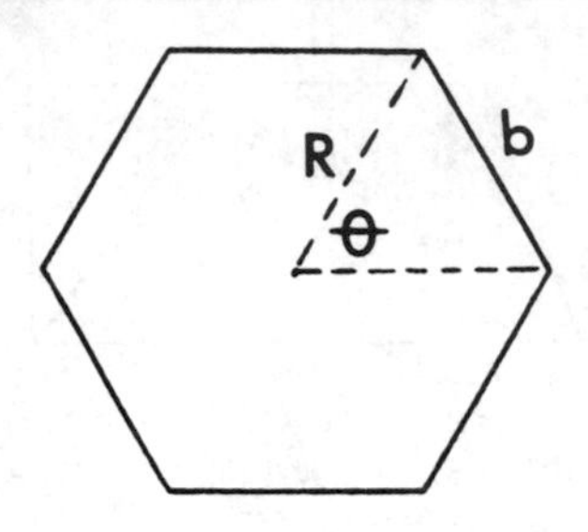

FIGURE 1.8. Regular polygon of n sides. $A = \frac{n}{4} b^2 \operatorname{ctn} \frac{\pi}{n}$; $R = \frac{b}{2} \csc \frac{\pi}{n}$.

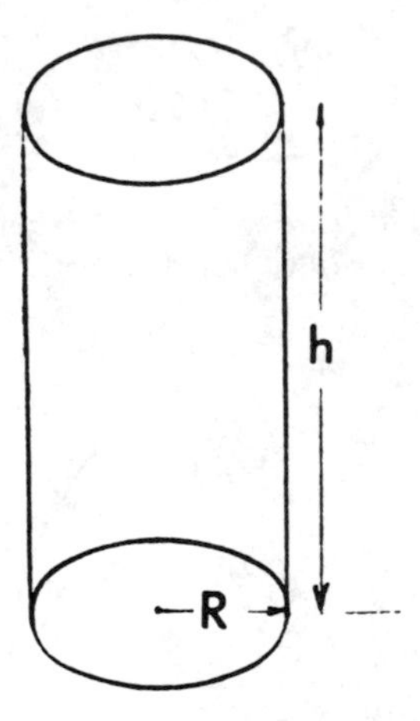

FIGURE 1.9. Right circular cylinder. $V = \pi R^2 h$; lateral surface area $= 2\pi Rh$.

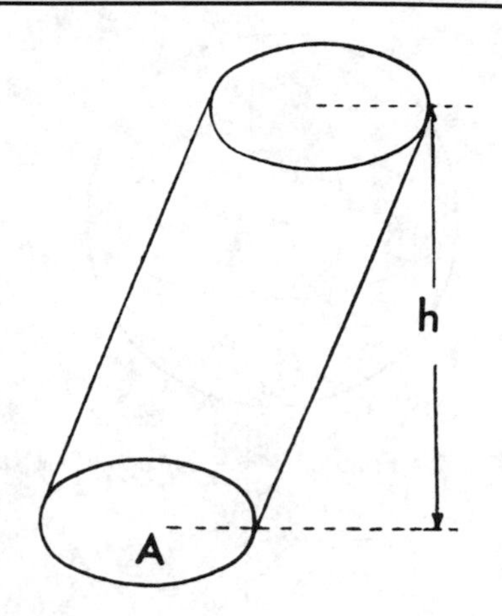

FIGURE 1.10. Cylinder (or prism) with parallel bases. $V = Ah$.

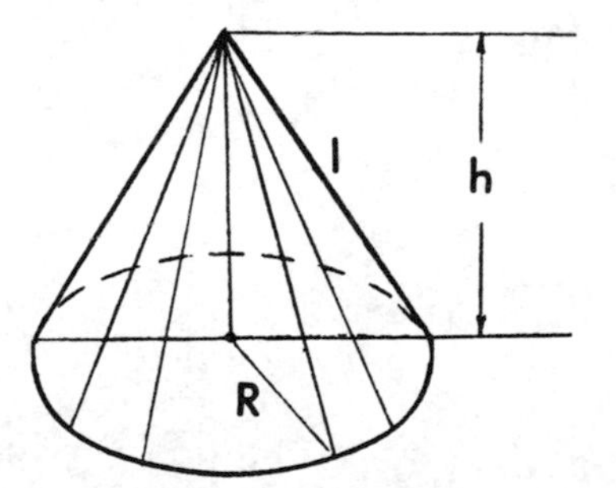

FIGURE 1.11. Right circular cone. $V = \frac{1}{3}\pi R^2 h$; lateral surface area $= \pi R l = \pi R\sqrt{R^2 + h^2}$.

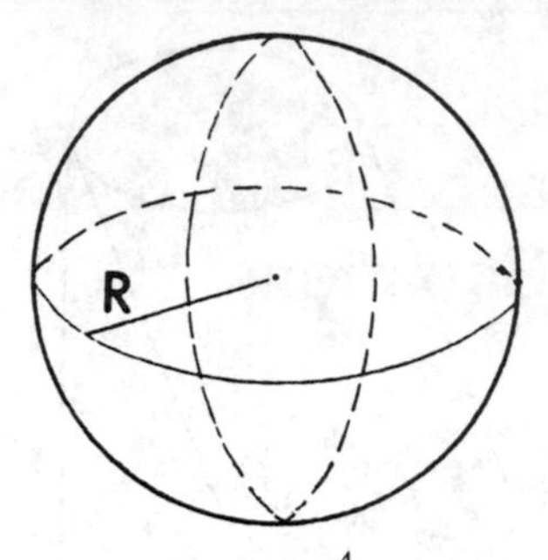

FIGURE 1.12. Sphere. $V = \frac{4}{3}\pi R^3$; surface area = $4\pi R^2$.

2 Determinants, Matrices, and Linear Systems of Equations

1. Determinants

Definition. The square array (matrix) A, with n rows and n columns, has associated with it the determinant

$$\det A = \begin{vmatrix} a_{11} & a_{12} & \cdots & a_{1n} \\ a_{21} & a_{22} & \cdots & a_{2n} \\ \cdots & \cdots & \cdots & \cdots \\ a_{n1} & a_{n2} & \cdots & a_{nn} \end{vmatrix},$$

a number equal to

$$\sum (\pm) a_{1i} a_{2j} a_{3k} \ldots a_{nl}$$

where $i, j, k, \ldots, l$ is a permutation of the n integers $1, 2, 3, \ldots, n$ in some order. The sign is plus if the permutation is *even* and is minus if the permutation is *odd* (see 1.12). The 2×2 determinant

$$\begin{vmatrix} a_{11} & a_{12} \\ a_{21} & a_{22} \end{vmatrix}$$

has the value $a_{11}a_{22} - a_{12}a_{21}$ since the permutation $(1, 2)$ is even and $(2, 1)$ is odd. For 3×3 determinants, permutations are as follows:

1,	2,	3	even
1,	3,	2	odd
2,	1,	3	odd
2,	3,	1	even
3,	1,	2	even
3,	2,	1	odd

Thus,

$$\begin{vmatrix} a_{11} & a_{12} & a_{13} \\ a_{21} & a_{22} & a_{23} \\ a_{31} & a_{32} & a_{33} \end{vmatrix} = \begin{Bmatrix} +a_{11} & \cdot & a_{22} & \cdot & a_{33} \\ -a_{11} & \cdot & a_{23} & \cdot & a_{32} \\ -a_{12} & \cdot & a_{21} & \cdot & a_{33} \\ +a_{12} & \cdot & a_{23} & \cdot & a_{31} \\ +a_{13} & \cdot & a_{21} & \cdot & a_{32} \\ -a_{13} & \cdot & a_{22} & \cdot & a_{31} \end{Bmatrix}$$

A determinant of order n is seen to be the sum of $n!$ signed products.

2. *Evaluation by Cofactors*

Each element a_{ij} has a determinant of order $(n-1)$ called a *minor* (M_{ij}) obtained by suppressing all elements in row i and column j. For example, the minor of element a_{22} in the 3×3 determinant above is

$$\begin{vmatrix} a_{11} & a_{13} \\ a_{31} & a_{33} \end{vmatrix}$$

The cofactor of element a_{ij}, denoted A_{ij}, is defined as $\pm M_{ij}$, where the sign is determined from i and j:

$$A_{ij} = (-1)^{i+j} M_{ij}.$$

The value of the $n \times n$ determinant equals the sum of products of elements of any row (or column) and their respective cofactors. Thus, for the 3×3 determinant

$$\det A = a_{11}A_{11} + a_{12}A_{12} + a_{13}A_{13} \text{ (first row)}$$

or

$$= a_{11}A_{11} + a_{21}A_{21} + a_{31}A_{31} \text{ (first column)}$$

etc.

3. *Properties of Determinants*

a. If the corresponding columns and rows of A are interchanged, det A is unchanged.

b. If any two rows (or columns) are interchanged, the sign of det A changes.

c. If any two rows (or columns) are identical, det $A = 0$.

d. If A is triangular (all elements above the main diagonal equal to zero), $A = a_{11} \cdot a_{22} \cdot \ldots \cdot a_{nn}$:

$$\begin{vmatrix} a_{11} & 0 & 0 & \cdots & 0 \\ a_{21} & a_{22} & 0 & \cdots & 0 \\ \cdots & \cdots & \cdots & \cdots & \cdots \\ a_{n1} & a_{n2} & a_{n3} & \cdots & a_{nn} \end{vmatrix}$$

e. If to each element of a row or column there is added C times the corresponding element in another row (or column), the value of the determinant is unchanged.

4. *Matrices*

Definition. A matrix is a rectangular array of numbers and is represented by a symbol A or $[a_{ij}]$:

$$A = \begin{bmatrix} a_{11} & a_{12} & \cdots & a_{1n} \\ a_{21} & a_{22} & \cdots & a_{2n} \\ \cdots & \cdots & \cdots & \cdots \\ a_{m1} & a_{m2} & \cdots & a_{mn} \end{bmatrix} = [a_{ij}]$$

The numbers a_{ij} are termed *elements* of the matrix; subscripts i and j identify the element as the number in row i and column j. The order of the matrix is $m \times n$ ("m by n"). When $m = n$, the matrix is square and is said to be of order n. For a square matrix of order n the elements $a_{11}, a_{22}, \ldots, a_{nn}$ constitute the main diagonal.

5. *Operations*

Addition. Matrices A and B of the same order may be added by adding corresponding elements, i.e., $A + B = [(a_{ij} + b_{ij})]$.

Scalar multiplication. If $A = [a_{ij}]$ and c is a constant (scalar), then $cA = [ca_{ij}]$, that is, every element of A is multiplied by c. In particular, $(-1)A = -A = [-a_{ij}]$ and $A + (-A) = 0$, a matrix with all elements equal to zero.

Multiplication of matrices. Matrices A and B may be multiplied only when they are conformable, which means that the number of columns of A equals the number of rows of B. Thus, if A is $m \times k$ and B is $k \times n$, then the product $C = AB$ exists as an $m \times n$ matrix with elements c_{ij} equal to the sum of products of elements in row i of A and corresponding elements of column j of B:

$$c_{ij} = \sum_{l=1}^{k} a_{il} b_{lj}$$

For example, if

$$\begin{bmatrix} a_{11} & a_{12} & \cdots & a_{1k} \\ a_{21} & a_{22} & \cdots & a_{2k} \\ \cdots & \cdots & \cdots & \cdots \\ a_{m1} & \cdots & \cdots & a_{mk} \end{bmatrix} \cdot \begin{bmatrix} b_{11} & b_{12} & \cdots & b_{1n} \\ b_{21} & b_{22} & \cdots & b_{2n} \\ \cdots & \cdots & \cdots & \cdots \\ b_{k1} & b_{k2} & \cdots & b_{kn} \end{bmatrix}$$

$$= \begin{bmatrix} c_{11} & c_{12} & \cdots & c_{1n} \\ c_{21} & c_{22} & \cdots & c_{2n} \\ \cdots & \cdots & \cdots & \\ c_{m1} & c_{m2} & \cdots & c_{mn} \end{bmatrix}$$

then element c_{21} is the sum of products $a_{21}b_{11} + a_{22}b_{21} + \ldots + a_{2k}b_{k1}$.

6. *Properties*

$$\begin{aligned} &A+B=B+A \\ &A+(B+C)=(A+B)+C \\ &(c_1+c_2)A=c_1A+c_2A \\ &c(A+B)=cA+cB \\ &c_1(c_2A)=(c_1c_2)A \\ &(AB)(C)=A(BC) \\ &(A+B)(C)=AC+BC \\ &AB \neq BA \text{ (in general)} \end{aligned}$$

7. *Transpose*

If A is an $n \times m$ matrix, the matrix of order $m \times n$ obtained by interchanging the rows and columns of A is called the *transpose* and is denoted A^T. The following are properties of A, B, and their respective transposes:

$$\begin{aligned}
(A^T)^T &= A \\
(A+B)^T &= A^T + B^T \\
(cA)^T &= cA^T \\
(AB)^T &= B^T A^T
\end{aligned}$$

A *symmetric* matrix is a square matrix A with the property $A = A^T$.

8. *Identity Matrix*

A square matrix in which each element of the main diagonal is the same constant a and all other elements zero is called a *scalar* matrix.

$$\begin{bmatrix}
a & 0 & 0 & \cdots & 0 \\
0 & a & 0 & \cdots & 0 \\
0 & 0 & a & \cdots & 0 \\
\cdots & \cdots & \cdots & \cdots & \\
0 & 0 & 0 & \cdots & a
\end{bmatrix}$$

When a scalar matrix multiplies a conformable second matrix A, the product is aA; that is, the same as multiplying A by a scalar a. A scalar matrix with diagonal elements 1 is called the *identity*, or *unit* matrix and is denoted I. Thus, for any nth order matrix A, the identity matrix of order n has the property

$$AI = IA = A$$

9. Adjoint

If A is an n-order square matrix and A_{ij} the cofactor of element a_{ij}, the transpose of $[A_{ij}]$ is called the *adjoint* of A:

$$adj A = [A_{ij}]^T$$

10. Inverse Matrix

Given a square matrix A of order n, if there exists a matrix B such that $AB = BA = I$, then B is called the *inverse* of A. The inverse is denoted A^{-1}. A necessary and sufficient condition that the square matrix A have an inverse is $\det A \neq 0$. Such a matrix is called *nonsingular*; its inverse is unique and it is given by

$$A^{-1} = \frac{adj A}{\det A}$$

Thus, to form the inverse of the nonsingular matrix A, form the adjoint of A and divide each element of the adjoint by $\det A$. For example,

$$\begin{bmatrix} 1 & 0 & 2 \\ 3 & -1 & 1 \\ 4 & 5 & 6 \end{bmatrix} \text{ has matrix of cofactors}$$

$$\begin{bmatrix} -11 & -14 & 19 \\ 10 & -2 & -5 \\ 2 & 5 & -1 \end{bmatrix},$$

$$\text{adjoint} = \begin{bmatrix} -11 & 10 & 2 \\ -14 & -2 & 5 \\ 19 & -5 & -1 \end{bmatrix} \text{ and determinant } 27.$$

Therefore,

$$A^{-1} = \begin{bmatrix} \frac{-11}{27} & \frac{10}{27} & \frac{2}{27} \\ \frac{-14}{27} & \frac{-2}{27} & \frac{5}{27} \\ \frac{19}{27} & \frac{-5}{27} & \frac{-1}{27} \end{bmatrix}.$$

11. Systems of Linear Equations

Given the system

$$\begin{array}{ccccccc} a_{11}x_1 & + & a_{12}x_2 & + \cdots + & a_{1n}x_n & = & b_1 \\ a_{21}x_1 & + & a_{22}x_2 & + \cdots + & a_{2n}x_n & = & b_2 \\ \vdots & & \vdots & \vdots & \vdots & & \vdots \\ a_{n1}x_1 & + & a_{n2}x_2 & + \cdots + & a_{nn}x_n & = & b_n \end{array}$$

a unique solution exists if $\det A \neq 0$, where A is the $n \times n$ matrix of coefficients $[a_{ij}]$.

- *Solution by Determinants (Cramer's Rule)*

$$x_1 = \begin{vmatrix} b_1 & a_{12} & \cdots & a_{1n} \\ b_2 & a_{22} & & \\ \vdots & \vdots & & \vdots \\ b_n & a_{n2} & & a_{nn} \end{vmatrix} \div \det A$$

$$x_2 = \begin{vmatrix} a_{11} & b_1 & a_{13} & \cdots & a_{1n} \\ a_{21} & b_2 & \cdots & & \cdots \\ \vdots & \vdots & & & \\ a_{n1} & b_n & a_{n3} & & a_{nn} \end{vmatrix} \div \det A$$

$$\vdots$$

$$x_k = \frac{\det A_k}{\det A},$$

where A_k is the matrix obtained from A by replacing the kth column of A by the column of b's.

12. *Matrix Solution*

The linear system may be written in matrix form $AX = B$ where A is the matrix of coefficients $[a_{ij}]$ and X and B are

$$X = \begin{bmatrix} x_1 \\ x_2 \\ \vdots \\ x_n \end{bmatrix} \qquad B = \begin{bmatrix} b_1 \\ b_2 \\ \vdots \\ b_n \end{bmatrix}$$

If a unique solution exists, $\det A \neq 0$; hence A^{-1} exists and

$$X = A^{-1}B.$$

3 Trigonometry

1. *Triangles*

In any triangle (in a plane) with sides a, b, and c and corresponding opposite angles A, B, C,

$$\frac{a}{\sin A} = \frac{b}{\sin B} = \frac{c}{\sin C}. \qquad \text{(Law of Sines)}$$

$$a^2 = b^2 + c^2 - 2cb\cos A. \qquad \text{(Law of Cosines)}$$

$$\frac{a+b}{a-b} = \frac{\tan\frac{1}{2}(A+B)}{\tan\frac{1}{2}(A-B)}. \qquad \text{(Law of Tangents)}$$

$$\sin\frac{1}{2}A = \sqrt{\frac{(s-b)(s-c)}{bc}}, \quad \text{where } \mathrm{s} = \frac{1}{2}(\mathrm{a}+\mathrm{b}+\mathrm{c}).$$

$$\cos\frac{1}{2}A = \sqrt{\frac{s(s-a)}{bc}}.$$

$$\tan\frac{1}{2}A = \sqrt{\frac{(s-b)(s-c)}{s(s-a)}}.$$

$$\begin{aligned}\text{Area} &= \frac{1}{2}bc\sin A \\ &= \sqrt{s(s-a)(s-b)(s-c)}.\end{aligned}$$

If the vertices have coordinates $(x_1, y_1), (x_2, y_2)$, (x_3, y_3), the area is the *absolute value* of the expression

$$\frac{1}{2}\begin{vmatrix} x_1 & y_1 & 1 \\ x_2 & y_2 & 1 \\ x_3 & y_3 & 1 \end{vmatrix}$$

2. *Trigonometric Functions of an Angle*

With reference to Figure 3.1, $P(x, y)$ is a point in either one of the four quadrants and A is an angle whose initial side is coincident with the positive x-axis and whose terminal side contains the point $P(x, y)$. The distance from the origin $P(x, y)$ is denoted by r and is positive. The trigonometric functions of the

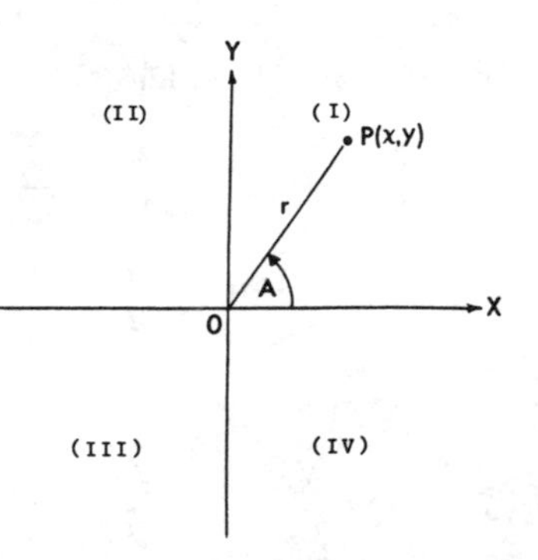

FIGURE 3.1. The trigonometric point. Angle A is taken to be positive when the rotation is counterclockwise and negative when the rotation is clockwise. The plane is divided into quadrants as shown.

angle A are defined as:

$$
\begin{aligned}
\sin A &= \text{sine } A &&= y/r \\
\cos A &= \text{cosine } A &&= x/r \\
\tan A &= \text{tangent } A &&= y/x \\
\text{ctn } A &= \text{cotangent } A &&= x/y \\
\sec A &= \text{secant } A &&= r/x \\
\csc A &= \text{cosecant } A &&= r/y
\end{aligned}
$$

Angles are measured in degrees or radians; $180° = \pi$ radians; 1 radian $= 180°/\pi$ degrees.

The trigonometric functions of 0°, 30°, 45°, and integer multiples of these are directly computed.

	0°	30°	45°	60°	90°	120°	135°	150°	180°
sin	0	$\frac{1}{2}$	$\frac{\sqrt{2}}{2}$	$\frac{\sqrt{3}}{2}$	1	$\frac{\sqrt{3}}{2}$	$\frac{\sqrt{2}}{2}$	$\frac{1}{2}$	0
cos	1	$\frac{\sqrt{3}}{2}$	$\frac{\sqrt{2}}{2}$	$\frac{1}{2}$	0	$-\frac{1}{2}$	$-\frac{\sqrt{2}}{2}$	$-\frac{\sqrt{3}}{2}$	-1
tan	0	$\frac{\sqrt{3}}{3}$	1	$\sqrt{3}$	∞	$-\sqrt{3}$	-1	$-\frac{\sqrt{3}}{3}$	0
ctn	∞	$\sqrt{3}$	1	$\frac{\sqrt{3}}{3}$	0	$-\frac{\sqrt{3}}{3}$	-1	$-\sqrt{3}$	∞
sec	1	$\frac{2\sqrt{3}}{3}$	$\sqrt{2}$	2	∞	-2	$-\sqrt{2}$	$-\frac{2\sqrt{3}}{3}$	-1
csc	∞	2	$\sqrt{2}$	$\frac{2\sqrt{3}}{3}$	1	$\frac{2\sqrt{3}}{3}$	$\sqrt{2}$	2	∞

3. *Trigonometric Identities*

$$\sin A = \frac{1}{\csc A}$$

$$\cos A = \frac{1}{\sec A}$$

$$\tan A = \frac{1}{\operatorname{ctn} A} = \frac{\sin A}{\cos A}$$

$$\csc A = \frac{1}{\sin A}$$

$$\sec A = \frac{1}{\cos A}$$

$$\operatorname{ctn} A = \frac{1}{\tan A} = \frac{\cos A}{\sin A}$$

$$\sin^2 A + \cos^2 A = 1$$

$$1 + \tan^2 A = \sec^2 A$$

$$1 + \operatorname{ctn}^2 A = \csc^2 A$$

$$\sin(A \pm B) = \sin A \cos B \pm \cos A \sin B$$

$$\cos(A \pm B) = \cos A \cos B \mp \sin A \sin B$$

$$\tan(A \pm B) = \frac{\tan A \pm \tan B}{1 \mp \tan A \tan B}$$

$$\sin 2A = 2 \sin A \cos A$$

$$\sin 3A = 3 \sin A - 4 \sin^3 A$$

$$\sin nA = 2\sin(n-1)A\cos A - \sin(n-2)A$$

$$\cos 2A = 2\cos^2 A - 1 = 1 - 2\sin^2 A$$

$$\cos 3A = 4\cos^3 A - 3\cos A$$

$$\cos nA = 2\cos(n-1)A\cos A - \cos(n-2)A$$

$$\sin A + \sin B = 2\sin\frac{1}{2}(A+B)\cos\frac{1}{2}(A-B)$$

$$\sin A - \sin B = 2\cos\frac{1}{2}(A+B)\sin\frac{1}{2}(A-B)$$

$$\cos A + \cos B = 2\cos\frac{1}{2}(A+B)\cos\frac{1}{2}(A-B)$$

$$\cos A - \cos B = -2\sin\frac{1}{2}(A+B)\sin\frac{1}{2}(A-B)$$

$$\tan A \pm \tan B = \frac{\sin(A \pm B)}{\cos A \cos B}$$

$$\text{ctn}\, A \pm \text{ctn}\, B = \pm\frac{\sin(A \pm B)}{\sin A \sin B}$$

$$\sin A \sin B = \frac{1}{2}\cos(A-B) - \frac{1}{2}\cos(A+B)$$

$$\cos A \cos B = \frac{1}{2}\cos(A-B) + \frac{1}{2}\cos(A+B)$$

$$\sin A \cos B = \frac{1}{2}\sin(A+B) + \frac{1}{2}\sin(A-B)$$

$$\sin\frac{A}{2} = \pm\sqrt{\frac{1-\cos A}{2}}$$

$$\cos\frac{A}{2} = \pm\sqrt{\frac{1+\cos A}{2}}$$

$$\tan\frac{A}{2} = \frac{1-\cos A}{\sin A} = \frac{\sin A}{1+\cos A} = \pm\sqrt{\frac{1-\cos A}{1+\cos A}}$$

$$\sin^2 A = \frac{1}{2}(1-\cos 2A)$$

$$\cos^2 A = \frac{1}{2}(1+\cos 2A)$$

$$\sin^3 A = \frac{1}{4}(3\sin A - \sin 3A)$$

$$\cos^3 A = \frac{1}{4}(\cos 3A + 3\cos A)$$

$$\sin ix = \frac{1}{2}i(e^x - e^{-x}) = i\sinh x$$

$$\cos ix = \frac{1}{2}(e^x + e^{-x}) = \cosh x$$

$$\tan ix = \frac{i(e^x - e^{-x})}{e^x + e^{-x}} = i\tanh x$$

$$e^{x+iy} = e^x(\cos y + i\sin y)$$

$$(\cos x \pm i\sin x)^n = \cos nx \pm i\sin nx$$

4. Inverse Trigonometric Functions

The inverse trigonometric functions are multiple valued, and this should be taken into account in the use of the following formulas.

$$\sin^{-1} x = \cos^{-1}\sqrt{1-x^2}$$

$$= \tan^{-1}\frac{x}{\sqrt{1-x^2}} = \operatorname{ctn}^{-1}\frac{\sqrt{1-x^2}}{x}$$

$$= \sec^{-1}\frac{1}{\sqrt{1-x^2}} = \csc^{-1}\frac{1}{x}$$

$$= -\sin^{-1}(-x)$$

$$\cos^{-1} x = \sin^{-1}\sqrt{1-x^2}$$

$$= \tan^{-1}\frac{\sqrt{1-x^2}}{x} = \operatorname{ctn}^{-1}\frac{x}{\sqrt{1-x^2}}$$

$$= \sec^{-1}\frac{1}{x} \qquad = \csc^{-1}\frac{1}{\sqrt{1-x^2}}$$

$$= \pi - \cos^{-1}(-x)$$

$$\tan^{-1} x = \operatorname{ctn}^{-1}\frac{1}{x}$$

$$= \sin^{-1}\frac{x}{\sqrt{1+x^2}} = \cos^{-1}\frac{1}{\sqrt{1+x^2}}$$

$$= \sec^{-1}\sqrt{1+x^2} = \csc^{-1}\frac{\sqrt{1+x^2}}{x}$$

$$= -\tan^{-1}(-x)$$

4 Analytic Geometry

1. Rectangular Coordinates

The points in a plane may be placed in one-to-one correspondence with pairs of real numbers. A common method is to use perpendicular lines that are horizontal and vertical and intersect at a point called the *origin*. These two lines constitute the coordinate axes; the horizontal line is the x-axis and the vertical line is the y-axis. The positive direction of the x-axis is to the right whereas the positive direction of the y-axis is up. If P is a point in the plane one may draw lines through it that are perpendicular to the x- and y-axes (such as the broken lines of Figure 4.1). The lines intersect the x-axis at a point with coordinate x_1 and the y-axis at a

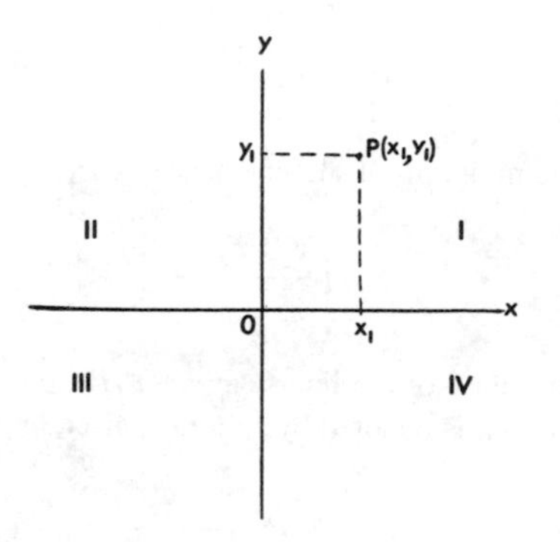

FIGURE 4.1. Rectangular coordinates.

point with coordinate y_1. We call x_1 the x-coordinate or *abscissa* and y_1 is termed the y-coordinate or *ordinate* of the point P. Thus, point P is associated with the pair of real numbers (x_1, y_1) and is denoted $P(x_1, y_1)$. The coordinate axes divide the plane into quadrants I, II, III, and IV.

2. *Distance between Two Points; Slope*

The distance d between the two points $P_1(x_1, y_1)$ and $P_2(x_2, y_2)$ is

$$d = \sqrt{(x_2 - x_1)^2 + (y_2 - y_1)^2}$$

In the special case when P_1 and P_2 are both on one of the coordinate axes, for instance, the x-axis,

$$d = \sqrt{(x_2 - x_1)^2} = |x_2 - x_1|,$$

or on the y-axis,

$$d = \sqrt{(y_2 - y_1)^2} = |y_2 - y_1|.$$

The midpoint of the line segment P_1P_2 is

$$\left(\frac{x_1 + x_2}{2}, \frac{y_1 + y_2}{2}\right).$$

The slope of the line segment P_1P_2, provided it is not vertical, is denoted by m and is given by

$$m = \frac{y_2 - y_1}{x_2 - x_1}.$$

The slope is related to the angle of inclination α (Figure 4.2) by

$$m = \tan \alpha$$

Two lines (or line segments) with slopes m_1 and m_2 are perpendicular if

$$m_1 = -1/m_2$$

and are parallel if $m_1 = m_2$.

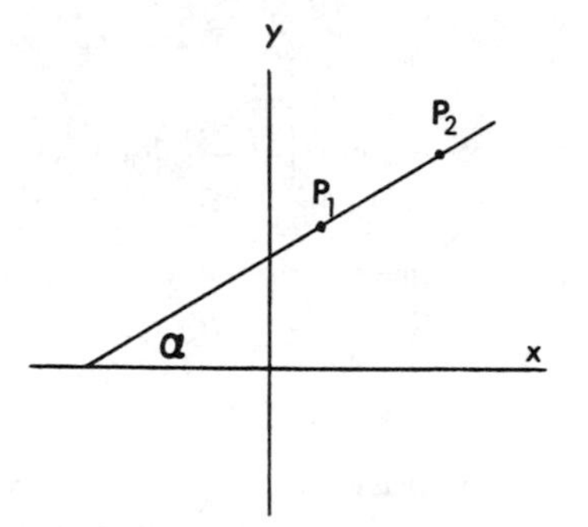

FIGURE 4.2. The angle of inclination is the smallest angle measured counterclockwise from the positive x-axis to the line that contains P_1P_2.

3. *Equations of Straight Lines*

A *vertical* line has an equation of the form

$$x = c$$

where $(c, 0)$ is its intersection with the x-axis. A line of slope m through point (x_1, y_1) is given by

$$y - y_1 = m(x - x_1)$$

Thus, a *horizontal line* (slope = 0) through point (x_1, y_1) is given by

$$y = y_1.$$

A nonvertical line through the two points $P_1(x_1, y_1)$ and $P_2(x_2, y_2)$ is given by either

$$y - y_1 = \left(\frac{y_2 - y_1}{x_2 - x_1}\right)(x - x_1)$$

or

$$y - y_2 = \left(\frac{y_2 - y_1}{x_2 - x_1}\right)(x - x_2).$$

A line with x-intercept a and y-intercept b is given by

$$\frac{x}{a} + \frac{y}{b} = 1 \qquad (a \neq 0, b \neq 0).$$

The *general equation* of a line is

$$Ax + By + C = 0$$

The *normal form* of the straight line equation is

$$x \cos \theta + y \sin \theta = p$$

where p is the distance along the normal from the origin and θ is the angle that the normal makes with the x-axis (Figure 4.3).

The general equation of the line $Ax + By + C = 0$ may be written in normal form by dividing by $\pm\sqrt{A^2 + B^2}$, where the plus sign is used when C is negative and the

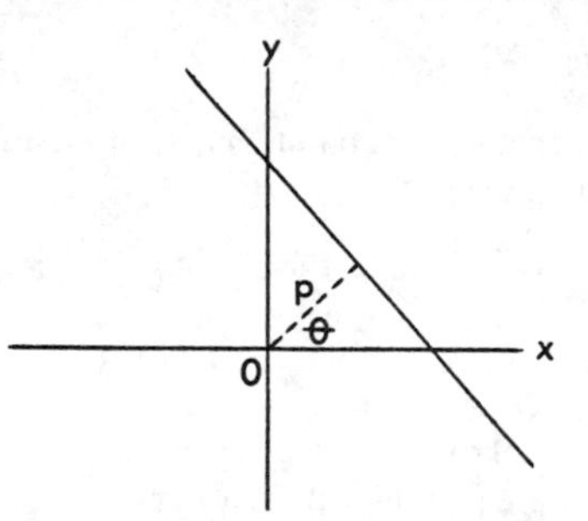

FIGURE 4.3. Construction for normal form of straight line equation.

minus sign is used when C is positive:

$$\frac{Ax+By+C}{\pm\sqrt{A^2+B^2}}=0,$$

so that

$$\cos\theta=\frac{A}{\pm\sqrt{A^2+B^2}},\qquad \sin\theta=\frac{B}{\pm\sqrt{A^2+B^2}}$$

and

$$p=\frac{|C|}{\sqrt{A^2+B^2}}.$$

4. *Distance from a Point to a Line*

The perpendicular distance from a point $P(x_1, y_1)$ to the line $Ax+By+C=0$ is given by d

$$d=\frac{Ax_1+By_1+C}{\pm\sqrt{A^2+B^2}}.$$

5. *Circle*

The general equation of a circle of radius r and center at $P(x_1, y_1)$ is

$$(x-x_1)^2+(y-y_1)^2=r^2.$$

6. *Parabola*

A parabola is the set of all points (x, y) in the plane that are equidistant from a given line called the *directrix* and a given point called the *focus*. The parabola is symmetric about a line that contains the focus and is perpendicular to the directrix. The line of symmetry intersects the parabola at its *vertex* (Figure 4.4). The eccentricity $e=1$.

The distance between the focus and the vertex, or vertex and directrix, is denoted by $p(>0)$ and leads to one of the following equations of a parabola with vertex at the origin (Figures 4.5 and 4.6):

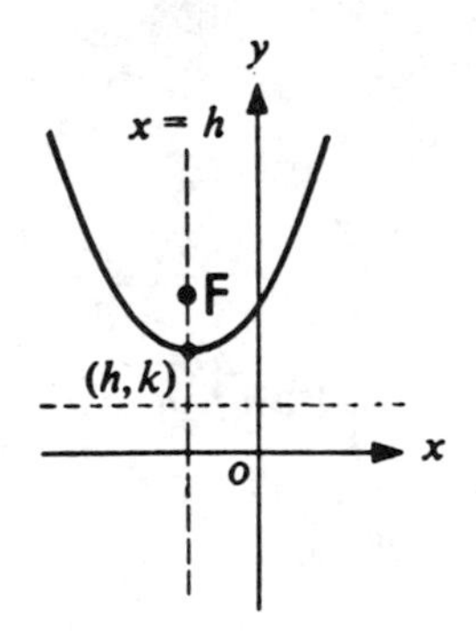

FIGURE 4.4. Parabola with vertex at (h, k). F identifies the focus.

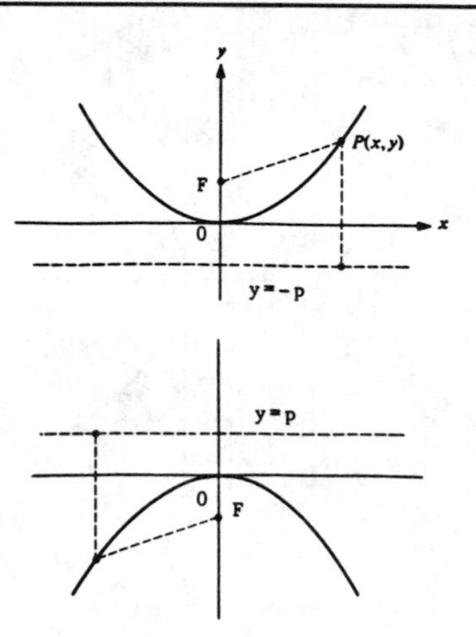

FIGURE 4.5. Parabolas with y-axis as the axis of symmetry and vertex at the origin. (Upper) $y = \frac{x^2}{4p}$; (lower) $y = -\frac{x^2}{4p}$.

$$y = \frac{x^2}{4p} \qquad \text{(opens upward)}$$

$$y = -\frac{x^2}{4p} \qquad \text{(opens downward)}$$

$$x = \frac{y^2}{4p} \qquad \text{(opens to right)}$$

$$x = -\frac{y^2}{4p} \qquad \text{(opens to left)}$$

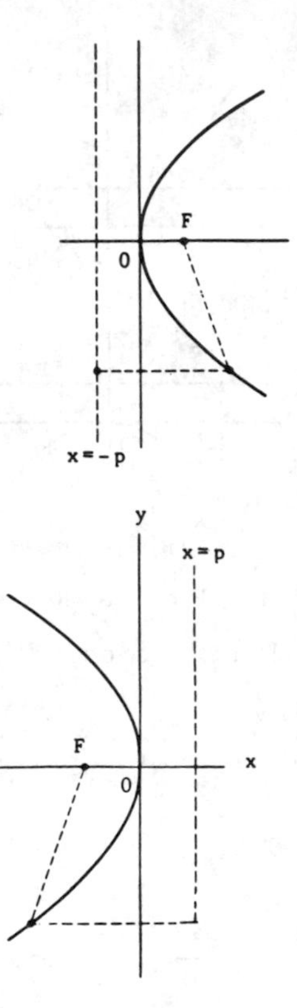

FIGURE 4.6. Parabolas with x-axis as the axis of symmetry and vertex at the origin. (Upper) $x = \frac{y^2}{4p}$; (lower) $x = -\frac{y^2}{4p}$.

For each of the four orientations shown in Figures 4.5 and 4.6, the coresponding parabola with vertex (h,k) is obtained by replacing x by $x-h$ and y by $y-k$. Thus, the parabola in Figure 4.7 has the equation

$$x-h=-\frac{(y-k)^2}{4p}.$$

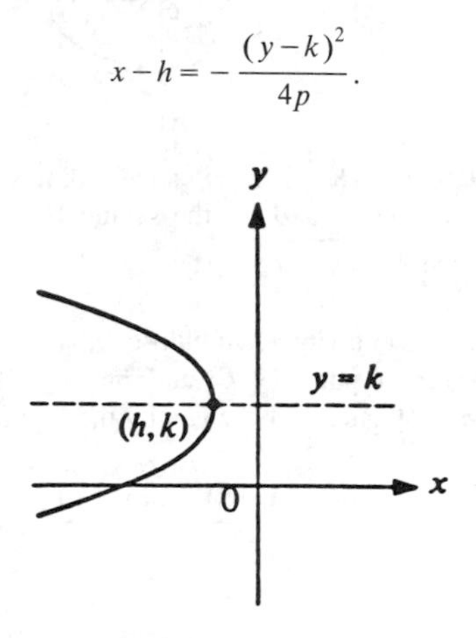

FIGURE 4.7. Parabola with vertex at (h,k) and axis parallel to the x-axis.

7. *Ellipse*

An ellipse is the set of all points in the plane such that the sum of their distances from two fixed points, called *foci*, is a given constant $2a$. The distance between the foci is denoted $2c$; the length of the major axis is $2a$, whereas the length of the minor axis is $2b$ (Figure 4.8) and

$$a=\sqrt{b^2+c^2}.$$

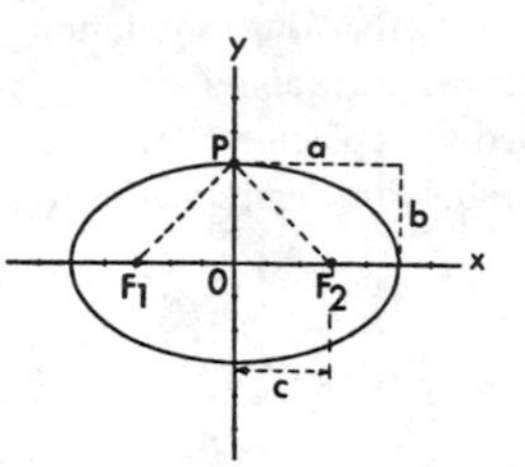

FIGURE 4.8. Ellipse; since point P is equidistant from foci F_1 and F_2 the segments F_1P and $F_2P = a$; hence $a = \sqrt{b^2 + c^2}$.

The eccentricity of an ellipse, e, is <1. An ellipse with center at point (h, k) and major axis *parallel to the x-axis* (Figure 4.9) is given by the equation

$$\frac{(x-h)^2}{a^2} + \frac{(y-k)^2}{b^2} = 1.$$

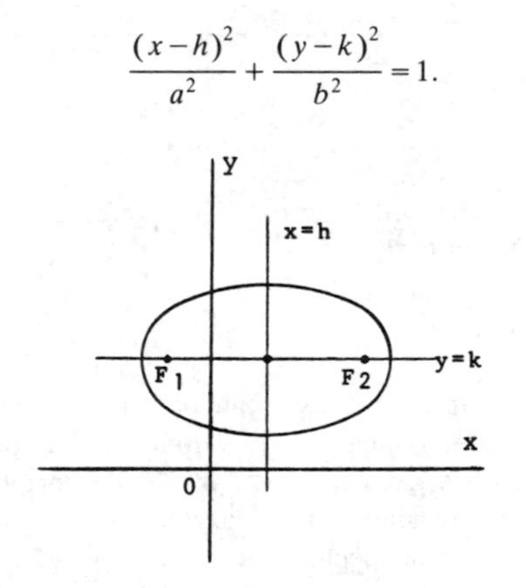

FIGURE 4.9. Ellipse with major axis parallel to the x-axis. F_1 and F_2 are the foci, each a distance c from center (h, k).

An ellipse with center at (h, k) and major axis parallel to the y-axis is given by the equation (Figure 4.10)

$$\frac{(y-k)^2}{a^2} + \frac{(x-h)^2}{b^2} = 1.$$

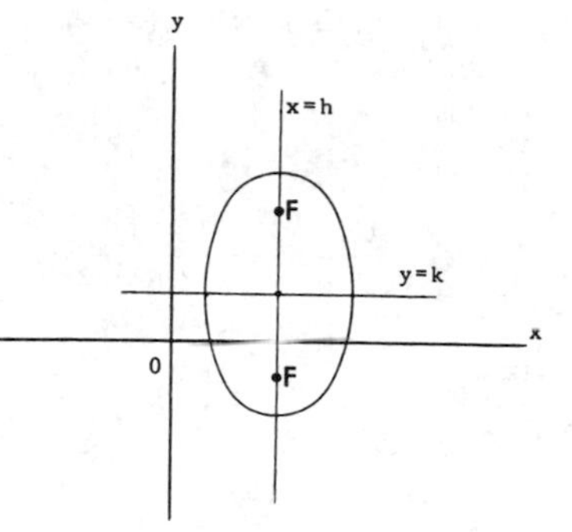

FIGURE 4.10. Ellipse with major axis parallel to the y-axis. Each focus is a distance c from center (h, k).

8. *Hyperbola ($e > 1$)*

A hyperbola is the set of all points in the plane such that the difference of its distances from two fixed points (foci) is a given positive constant denoted $2a$. The distance between the two foci is $2c$ and that between the two vertices is $2a$. The quantity b is defined by the equation

$$b = \sqrt{c^2 - a^2}$$

and is illustrated in Figure 4.11, which shows the construction of a hyperbola given by the equation

$$\frac{x^2}{a^2} - \frac{y^2}{b^2} = 1.$$

When the focal axis is parallel to the y-axis the equation of the hyperbola with center (h, k) (Figures 4.12 and 4.13) is

$$\frac{(y-k)^2}{a^2} - \frac{(x-h)^2}{b^2} = 1.$$

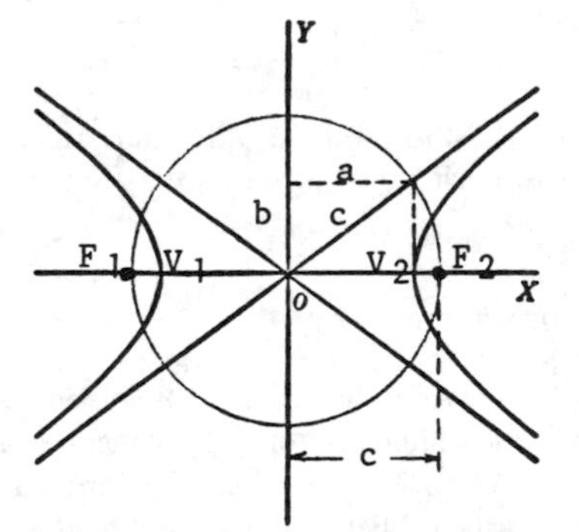

FIGURE 4.11. Hyperbola; V_1, V_2 = vertices; F_1, F_2 = foci. A circle at center 0 with radius c contains the vertices and illustrates the relation among a, b, and c. Asymptotes have slopes b/a and $-b/a$ for the orientation shown.

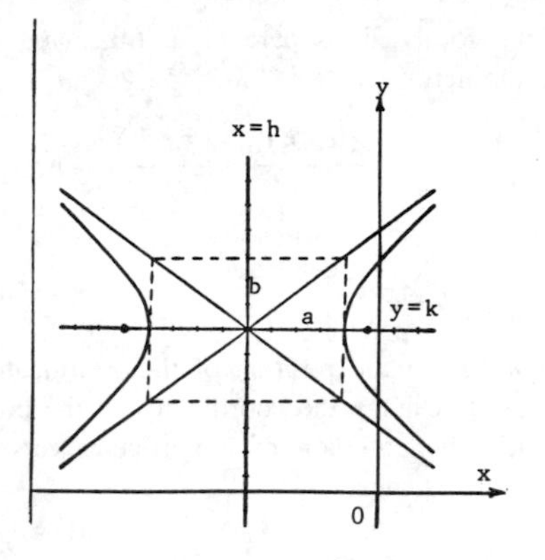

FIGURE 4.12. Hyperbola with center at (h, k): $\dfrac{(x-h)^2}{a^2} - \dfrac{(y-k)^2}{b^2} = 1$; slopes of asymptotes $\pm b/a$.

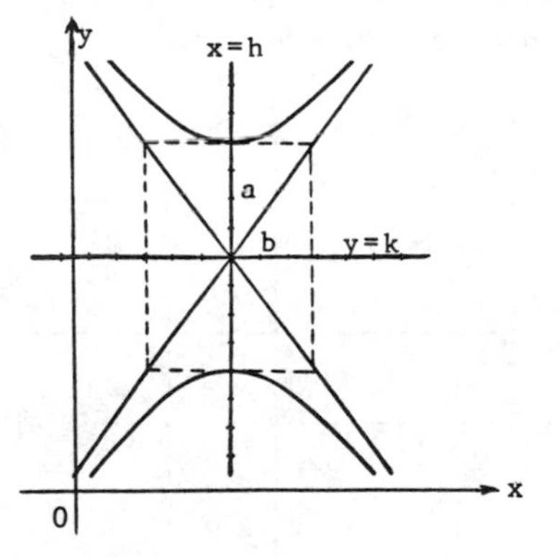

FIGURE 4.13. Hyperbola with center at (h, k): $\dfrac{(y-k)^2}{a^2} - \dfrac{(x-h)^2}{b^2} = 1$; slopes of asymptotes $\pm a/b$.

If the focal axis is parallel to the x-axis and center (h,k), then

$$\frac{(x-h)^2}{a^2}-\frac{(y-k)^2}{b^2}=1$$

9. *Change of Axes*

A change in the position of the coordinate axes will generally change the coordinates of the points in the plane. The equation of a particular curve will also generally change.

- *Translation*

When the new axes remain parallel to the original, the transformation is called a *translation* (Figure 4.14). The new axes, denoted x' and y', have origin $0'$ at (h,k) with reference to the x and y axes.

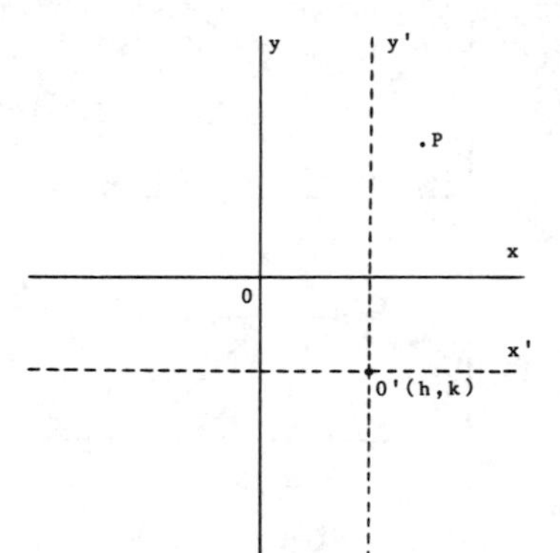

FIGURE 4.14. Translation of axes.

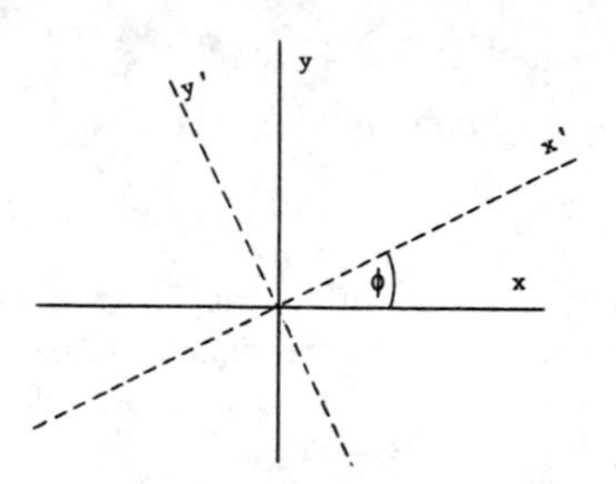

FIGURE 4.15. Rotation of axes.

A point P with coordinates (x, y) with respect to the original has coordinates (x', y') with respect to the new axes. These are related by

$$x = x' + h$$

$$y = y' + k$$

For example, the ellipse of Figure 4.10 has the following simpler equation with respect to axes x' and y' with the center at (h, k):

$$\frac{y'^2}{a^2} + \frac{x'^2}{b^2} = 1.$$

- *Rotation*

When the new axes are drawn through the same origin, remaining mutually perpendicular, but tilted with respect to the original, the transformation is one of rotation. For angle of rotation ϕ (Figure 4.15), the coordinates (x, y) and (x', y') of a point P are related by

$$x = x' \cos \phi - y' \sin \phi$$

$$y = x' \sin \phi + y' \cos \phi$$

10. *General Equation of Degree Two*

$$Ax^2 + Bxy + Cy^2 + Dx + Ey + F = 0$$

Every equation of the above form defines a conic section or one of the limiting forms of a conic. By rotating the axes through a particular angle ϕ, the xy-term vanishes, yielding

$$A'x'^2 + C'y'^2 + D'x' + E'y' + F' = 0$$

with respect to the axes x' and y'. The required angle ϕ (see Figure 4.15) is calculated from

$$\tan 2\phi = \frac{B}{A - C}, \qquad (\phi < 90°).$$

11. *Polar Coordinates (Figure 4.16)*

The fixed point O is the origin or *pole* and a line *OA* drawn through it is the polar axis. A point *P* in the plane is determined from its distance *r*, measured from

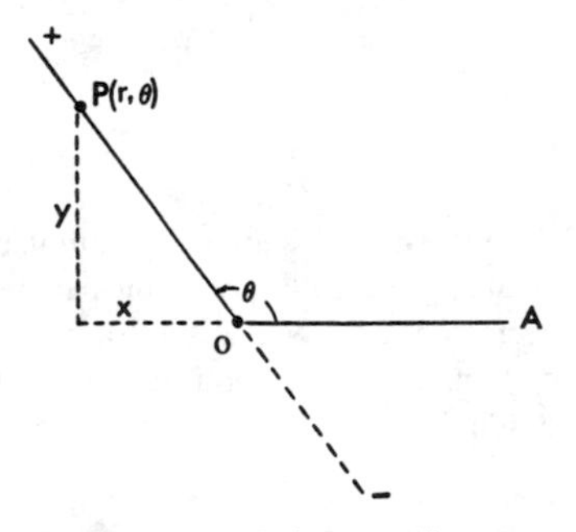

FIGURE 4.16. Polar coordinates.

O, and the angle θ between OP and OA. Distances measured on the terminal line of θ from the pole are positive, whereas those measured in the opposite direction are negative.

Rectangular coordinates (x, y) and polar coordinates (r, θ) are related according to

$$x = r\cos\theta, \qquad y = r\sin\theta$$

$$r^2 = x^2 + y^2, \qquad \tan\theta = y/x.$$

Several well-known polar curves are shown in Figures 4.17 to 4.21.

The polar equation of a conic section with focus at the pole and distance $2p$ from directrix to focus is either

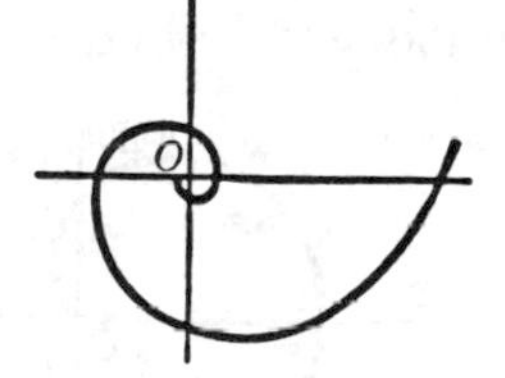

FIGURE 4.17. Polar curve $r = e^{a\theta}$.

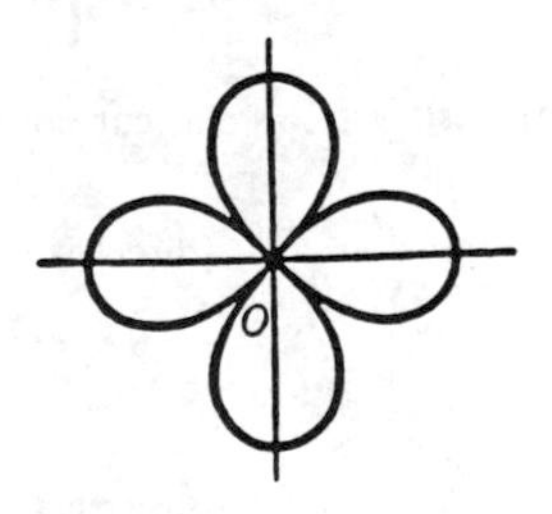

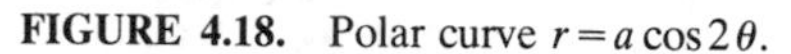

FIGURE 4.18. Polar curve $r = a\cos 2\theta$.

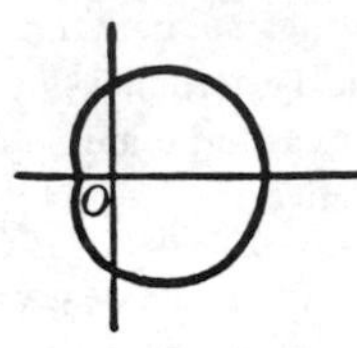

FIGURE 4.19. Polar curve $r = 2a\cos\theta + b$.

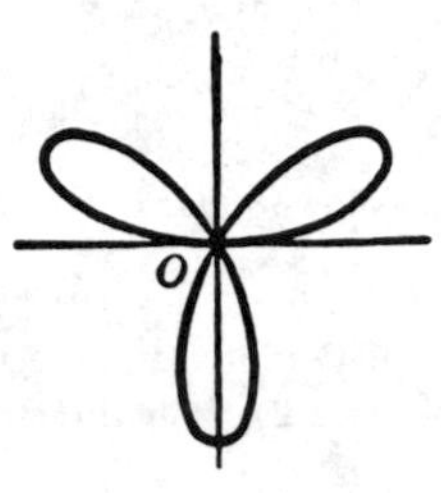

FIGURE 4.20. Polar curve $r = a\sin 3\theta$.

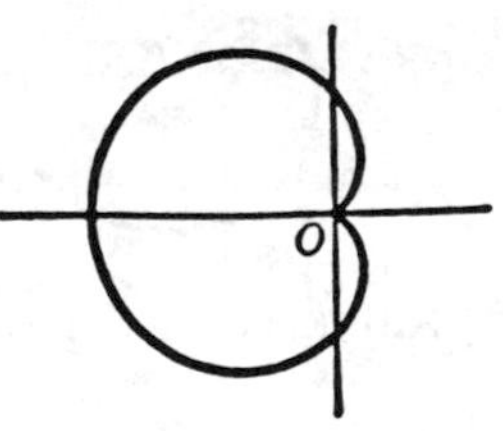

FIGURE 4.21. Polar curve $r = a(1 - \cos\theta)$.

$$r = \frac{2ep}{1 - e\cos\theta} \qquad \text{(directrix to left of pole)}$$

or

$$r = \frac{2ep}{1 + e\cos\theta} \qquad \text{(directrix to right of pole)}$$

The corresponding equations for the directrix below or above the pole are as above, except that $\sin\,\theta$ appears instead of $\cos\,\theta$.

12. Curves and Equations

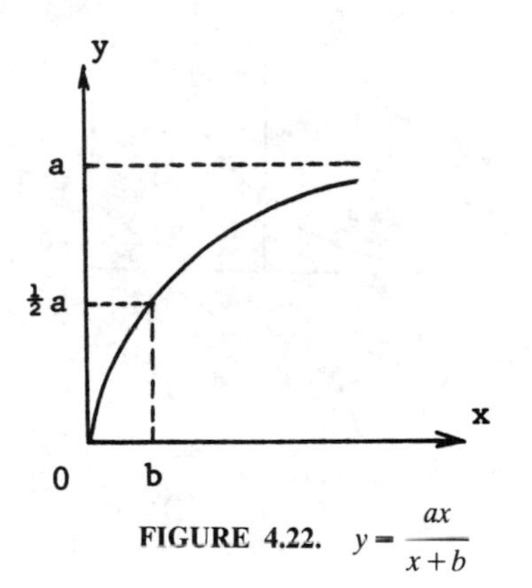

FIGURE 4.22. $y = \dfrac{ax}{x+b}$

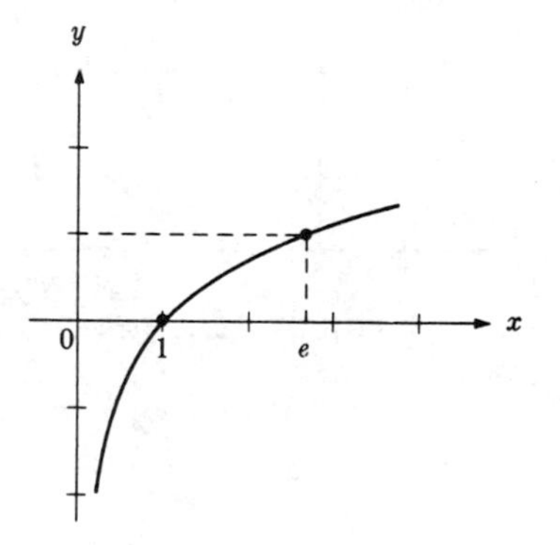

FIGURE 4.23. $y = \log x.$

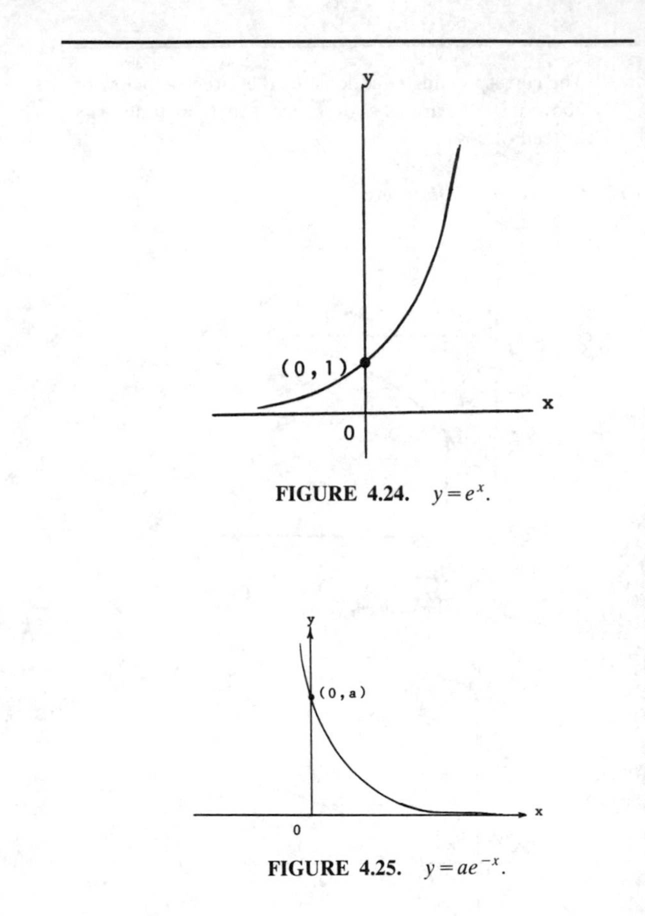

FIGURE 4.24. $y = e^x$.

FIGURE 4.25. $y = ae^{-x}$.

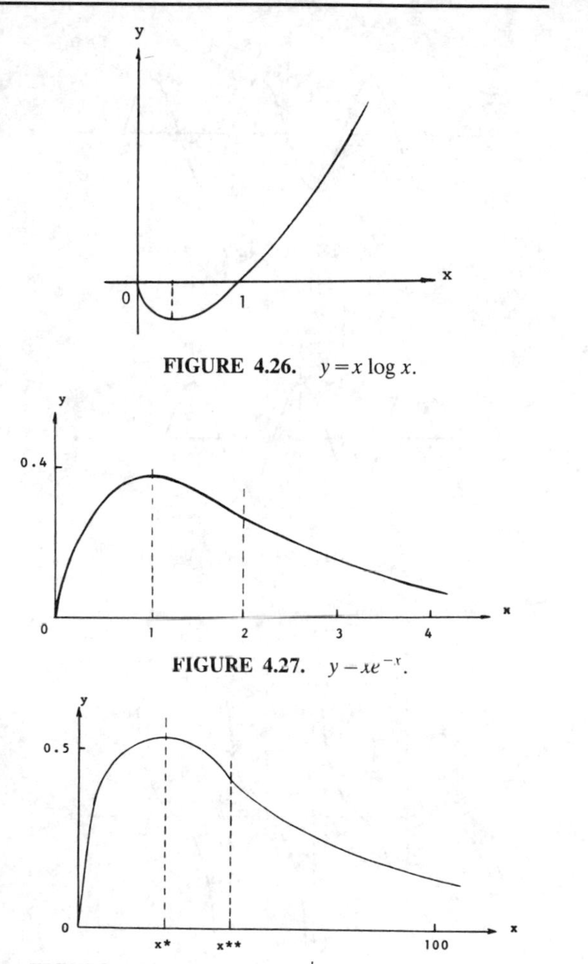

FIGURE 4.26. $y = x \log x$.

FIGURE 4.27. $y = xe^{-x}$.

FIGURE 4.28. $y = e^{-ax} - e^{-bx}$, $0 < a < b$ (drawn for $a = 0.02$, $b = 0.1$, and showing maximum and inflection).

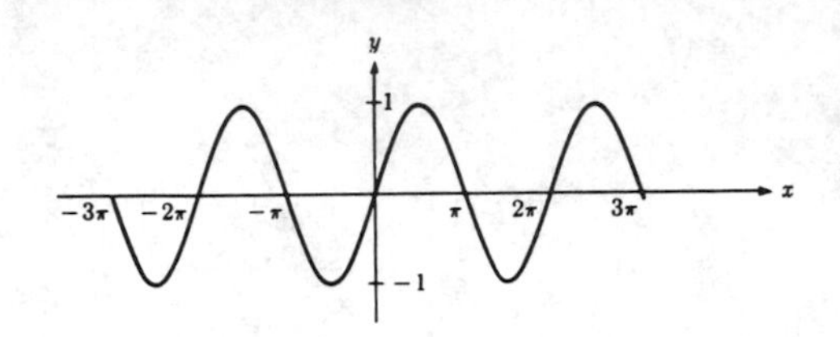

FIGURE 4.29. $y = \sin x$.

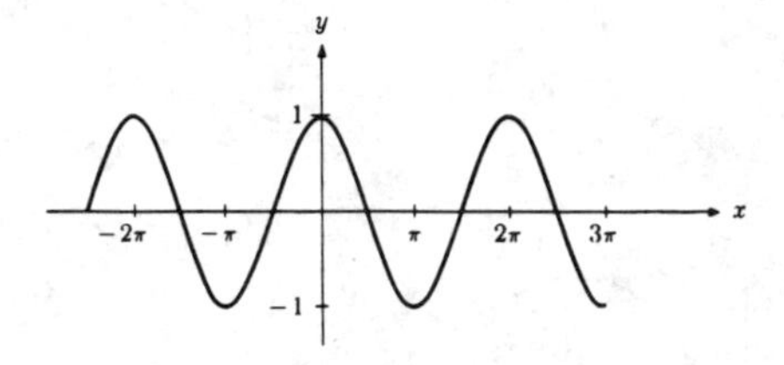

FIGURE 4.30. $y = \cos x$.

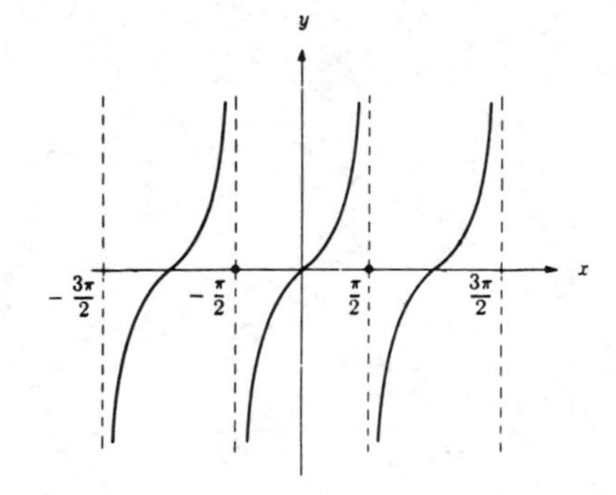

FIGURE 4.31. $y = \tan x$.

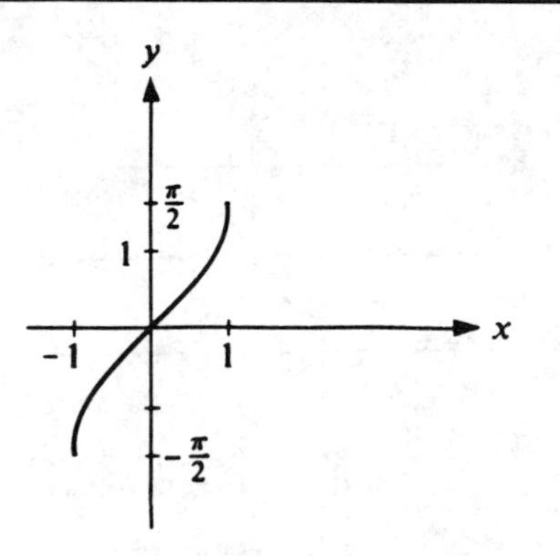

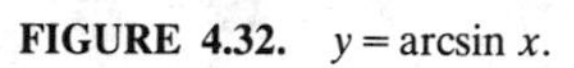

FIGURE 4.32. $y = \arcsin x$.

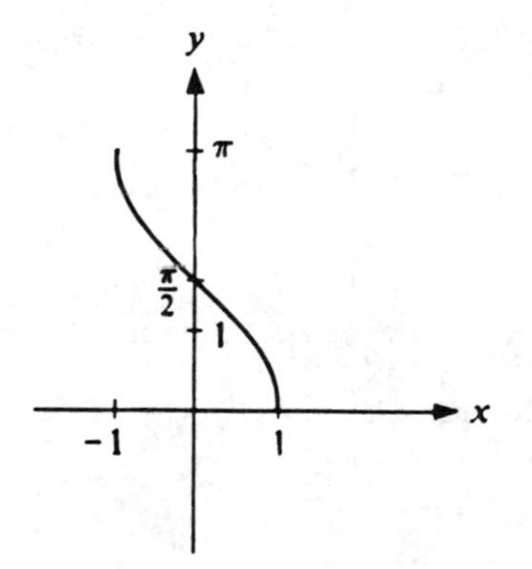

FIGURE 4.33. $y = \arccos x$.

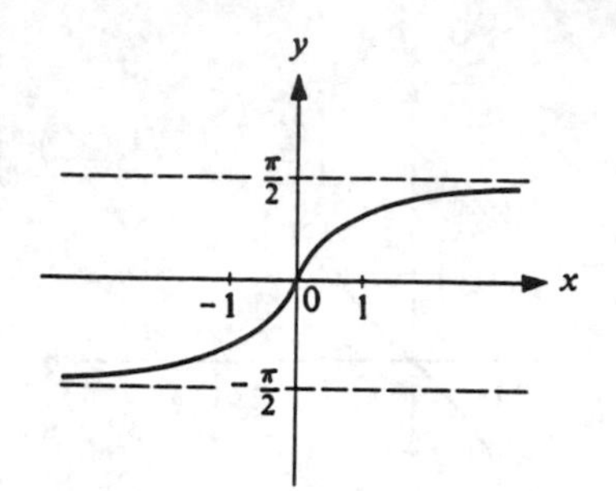

FIGURE 4.34. $y = \arctan x$.

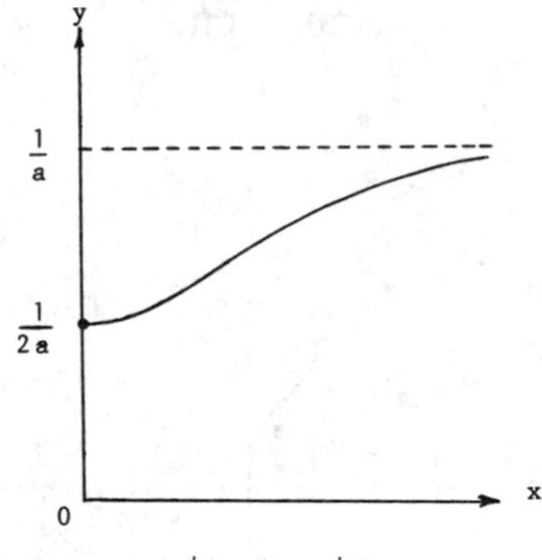

FIGURE 4.35. $y = e^{bx}/a(1 + e^{bx})$, $x \geq 0$ (logistic equation).

5 Series

1. Bernoulli and Euler Numbers

A set of numbers, $B_1, B_3, \ldots, B_{2n-1}$ (Bernoulli numbers) and $B_2, B_4, \ldots, B_{2n}$ (Euler numbers) appear in the series expansions of many functions. A partial listing follows; these are computed from the following equations:

$$B_{2n} - \frac{2n(2n-1)}{2!} B_{2n-2} + \frac{2n(2n-1)(2n-2)(2n-3)}{4!} B_{2n-4} - \ldots + (-1)^n = 0,$$

and

$$\frac{2^{2n}(2^{2n}-1)}{2n} B_{2n-1} = (2n-1) B_{2n-2} - \frac{(2n-1)(2n-2)(2n-3)}{3!} B_{2n-4} + \ldots + (-1)^{n-1}.$$

$$\begin{aligned} B_1 &= 1/6 & B_2 &= 1 \\ B_3 &= 1/30 & B_4 &= 5 \\ B_5 &= 1/42 & B_6 &= 61 \\ B_7 &= 1/30 & B_8 &= 1385 \\ B_9 &= 5/66 & B_{10} &= 50521 \end{aligned}$$

$$B_{11} = 691/2730 \qquad B_{12} = 2702765$$
$$B_{13} = 7/6 \qquad B_{14} = 199360981$$
$$\vdots \qquad \vdots$$

2. *Series of Functions*

In the following, the interval of convergence is indicated, otherwise it is all x. Logarithms are to the base e. Bernoulli and Euler numbers (B_{2n-1} and B_{2n}) appear in certain expressions.

$$(a+x)^n = a^n + na^{n-1}x + \frac{n(n-1)}{2!}a^{n-2}x^2$$
$$+ \frac{n(n-1)(n-2)}{3!}a^{n-3}x^3 + \dots$$
$$+ \frac{n!}{(n-j)!j!}a^{n-j}x^j + \dots \quad [x^2 < a^2]$$

$$(a-bx)^{-1} = \frac{1}{a}\left[1 + \frac{bx}{a} + \frac{b^2x^2}{a^2} + \frac{b^3x^3}{a^3} + \dots\right]$$
$$[b^2x^2 < a^2]$$

$$(1 \pm x)^n = 1 \pm nx + \frac{n(n-1)}{2!}x^2$$
$$\pm \frac{n(n-1)(n-2)x^3}{3!} + \dots \quad [x^2 < 1]$$

$$(1 \pm x)^{-n} = 1 \mp nx + \frac{n(n+1)}{2!}x^2$$
$$\mp \frac{n(n+1)(n+2)}{3!}x^3 + \dots \quad [x^2 < 1]$$

$$(1 \pm x)^{\frac{1}{2}} = 1 \pm \frac{1}{2}x - \frac{1}{2 \cdot 4}x^2 \pm \frac{1 \cdot 3}{2 \cdot 4 \cdot 6}x^3 - \frac{1 \cdot 3 \cdot 5}{2 \cdot 4 \cdot 6 \cdot 8}x^4 \pm \ldots \qquad [x^2 < 1]$$

$$(1 \pm x)^{-\frac{1}{2}} = 1 \mp \frac{1}{2}x + \frac{1 \cdot 3}{2 \cdot 4}x^2 \mp \frac{1 \cdot 3 \cdot 5}{2 \cdot 4 \cdot 6}x^3 + \frac{1 \cdot 3 \cdot 5 \cdot 7}{2 \cdot 4 \cdot 6 \cdot 8}x^4 \mp \ldots \qquad [x^2 < 1]$$

$$(1 \pm x^2)^{\frac{1}{2}} = 1 \pm \frac{1}{2}x^2 - \frac{x^4}{2 \cdot 4} \pm \frac{1 \cdot 3}{2 \cdot 4 \cdot 6}x^6 - \frac{1 \cdot 3 \cdot 5}{2 \cdot 4 \cdot 6 \cdot 8}x^8 \pm \ldots \qquad [x^2 < 1]$$

$$(1 \pm x)^{-1} = 1 \mp x + x^2 \mp x^3 + x^4 \mp x^5 + \ldots \qquad [x^2 < 1]$$

$$(1 \pm x)^{-2} = 1 \mp 2x + 3x^2 \mp 4x^3 + 5x^4 \mp \ldots \qquad [x^2 < 1]$$

$$e^x = 1 + x + \frac{x^2}{2!} + \frac{x^3}{3!} + \frac{x^4}{4!} + \ldots$$

$$e^{-x^2} = 1 - x^2 + \frac{x^4}{2!} - \frac{x^6}{3!} + \frac{x^8}{4!} - \ldots$$

$$a^x = 1 + x \log a + \frac{(x \log a)^2}{2!} + \frac{(x \log a)^3}{3!} + \ldots$$

$$\log x = (x-1) - \frac{1}{2}(x-1)^2 + \frac{1}{3}(x-1)^3 - \ldots$$

$$[0 < x < 2]$$

$$\log x = \frac{x-1}{x} + \frac{1}{2}\left(\frac{x-1}{x}\right)^2 + \frac{1}{3}\left(\frac{x-1}{x}\right)^3 + \ldots$$

$$\left[x > \frac{1}{2}\right]$$

$$\log x = 2\left[\left(\frac{x-1}{x+1}\right) + \frac{1}{3}\left(\frac{x-1}{x+1}\right)^3 + \frac{1}{5}\left(\frac{x-1}{x+1}\right)^5 + \ldots\right]$$

$$[x > 0]$$

$$\log(1+x) = x - \frac{1}{2}x^2 + \frac{1}{3}x^3 - \frac{1}{4}x^4 + \ldots$$

$$[x^2 < 1]$$

$$\log\left(\frac{1+x}{1-x}\right) = 2\left[x + \frac{1}{3}x^3 + \frac{1}{5}x^5 + \frac{1}{7}x^7 + \ldots\right]$$

$$[x^2 < 1]$$

$$\log\left(\frac{x+1}{x-1}\right) = 2\left[\frac{1}{x} + \frac{1}{3}\left(\frac{1}{x}\right)^3 + \frac{1}{5}\left(\frac{1}{x}\right)^5 + \ldots\right]$$

$$[x^2 > 1]$$

$$\sin x = x - \frac{x^3}{3!} + \frac{x^5}{5!} - \frac{x^7}{7!} + \ldots$$

$$\cos x = 1 - \frac{x^2}{2!} + \frac{x^4}{4!} - \frac{x^6}{6!} + \ldots$$

$$\tan x = x + \frac{x^3}{3} + \frac{2x^5}{15} + \frac{17x^7}{315}$$

$$+ \ldots + \frac{2^{2n}(2^{2n}-1)B_{2n-1}x^{2n-1}}{(2n)!}$$

$$\left[x^2 < \frac{\pi^2}{4}\right]$$

$$\text{ctn } x = \frac{1}{x} - \frac{x}{3} - \frac{x^3}{45} - \frac{2x^5}{945}$$

$$- \ldots - \frac{B_{2n-1}(2x)^{2n}}{(2n)!x} - \ldots$$

$$[x^2 < \pi^2]$$

$$\sec x = 1 + \frac{x^2}{2!} + \frac{5x^4}{4!} + \frac{61x^6}{6!} + \ldots$$

$$+ \frac{B_{2n}x^{2n}}{(2n)!} + \ldots \qquad \left[x^2 < \frac{\pi^2}{4}\right]$$

$$\csc x = \frac{1}{x} + \frac{x}{3!} + \frac{7x^3}{3 \cdot 5!} + \frac{31x^5}{3 \cdot 7!}$$

$$+ \ldots + \frac{2(2^{2n+1}-1)}{(2n+2)!}B_{2n+1}x^{2n+1} + \ldots$$

$$[x^2 < \pi^2]$$

$$\sin^{-1} x = x + \frac{x^3}{6} + \frac{(1 \cdot 3)x^5}{(2 \cdot 4)5} + \frac{(1 \cdot 3 \cdot 5)x^7}{(2 \cdot 4 \cdot 6)7} + \ldots$$

$$[x^2 < 1]$$

$$\tan^{-1} x = x - \frac{1}{3}x^3 + \frac{1}{5}x^5 - \frac{1}{7}x^7 + \ldots$$

$$[x^2 < 1]$$

$$\sec^{-1} x = \frac{\pi}{2} - \frac{1}{x} - \frac{1}{6x^3}$$

$$- \frac{1 \cdot 3}{(2 \cdot 4)5x^5} - \frac{1 \cdot 3 \cdot 5}{(2 \cdot 4 \cdot 6)7x^7} - \ldots$$

$$[x^2 > 1]$$

$$\sinh x = x + \frac{x^3}{3!} + \frac{x^5}{5!} + \frac{x^7}{7!} + \ldots$$

$$\cosh x = 1 + \frac{x^2}{2!} + \frac{x^4}{4!} + \frac{x^6}{6!} + \frac{x^8}{8!} + \ldots$$

$$\tanh x = (2^2 - 1)2^2 B_1 \frac{x}{2!} - (2^4 - 1)2^4 B_3 \frac{x^3}{4!}$$

$$+ (2^6 - 1)2^6 B_5 \frac{x^5}{6!} - \ldots \qquad \left[x^2 < \frac{\pi^2}{4}\right]$$

$$\operatorname{ctnh} x = \frac{1}{x}\left(1 + \frac{2^2 B_1 x^2}{2!} - \frac{2^4 B_3 x^4}{4!}\right.$$

$$\left. + \frac{2^6 B_5 x^6}{6!} - \ldots\right)$$

$$[x^2 < \pi^2]$$

$$\operatorname{sech} x = 1 - \frac{B_2 x^2}{2!} + \frac{B_4 x^4}{4!} - \frac{B_6 x^6}{6!} + \dots$$

$$\left[x^2 < \frac{\pi^2}{4}\right]$$

$$\operatorname{csch} x = \frac{1}{x} - (2-1)2B_1 \frac{x}{2!}$$

$$+ (2^3 - 1)2B_3 \frac{x^3}{4!} - \dots$$

$$[x^2 < \pi^2]$$

$$\sinh^{-1} x = x - \frac{1}{2}\frac{x^3}{3} + \frac{1\cdot 3}{2\cdot 4}\frac{x^5}{5} - \frac{1\cdot 3\cdot 5}{2\cdot 4\cdot 6}\frac{x^7}{7} + \dots$$

$$[x^2 < 1]$$

$$\tanh^{-1} x = x + \frac{x^3}{3} + \frac{x^5}{5} + \frac{x^7}{7} + \dots \qquad [x^2 < 1]$$

$$\operatorname{ctnh}^{-1} x = \frac{1}{x} + \frac{1}{3x^3} + \frac{1}{5x^5} + \dots \qquad [x^2 > 1]$$

$$\operatorname{csch}^{-1} x = \frac{1}{x} - \frac{1}{2\cdot 3x^3} + \frac{1\cdot 3}{2\cdot 4\cdot 5x^5}$$

$$- \frac{1\cdot 3\cdot 5}{2\cdot 4\cdot 6\cdot 7x^7} + \dots \qquad [x^2 > 1]$$

$$\int_0^x e^{-t^2}\,dt = x - \frac{1}{3}x^3 + \frac{x^5}{5\cdot 2!} - \frac{x^7}{7\cdot 3!} + \dots$$

3. *Error Function*

The following function, known as the error function, erf x, arises frequently in applications:

$$\text{erf}\, x = \frac{2}{\sqrt{\pi}} \int_0^x e^{-t^2}\, dt$$

The integral cannot be represented in terms of a finite number of elementary functions, therefore values of erf x have been compiled in tables. The following is the series for erf x:

$$\text{erf}\, x = \frac{2}{\sqrt{\pi}} \left[x - \frac{x^3}{3} + \frac{x^5}{5 \cdot 2!} - \frac{x^7}{7 \cdot 3!} + \ldots \right]$$

There is a close relation between this function and the area under the standard normal curve (Table A.1). For evaluation it is convenient to use z instead of x; then erf z may be evaluated from the area $F(z)$ given in Table A.1 by use of the relation

$$\text{erf}\, z = 2F(\sqrt{2}\, z)$$

Example

$$\text{erf}(0.5) = 2F[(1.414)(0.5)] = 2F(0.707)$$

By interpolation from Table A.1, $F(0.707) = 0.260$; thus, $\text{erf}(0.5) = 0.520$.

6 Differential Calculus

1. *Notation*

For the following equations, the symbols $f(x)$, $g(x)$, etc., represent functions of x. The value of a function $f(x)$ at $x=a$ is denoted $f(a)$. For the function $y=f(x)$ the derivative of y with respect to x is denoted by one of the following:

$$\frac{dy}{dx}, \quad f'(x), \quad D_x y, \quad y'.$$

Higher derivatives are as follows:

$$\frac{d^2y}{dx^2}=\frac{d}{dx}\left(\frac{dy}{dx}\right)=\frac{d}{dx}f'(x)=f''(x)$$

$$\frac{d^3y}{dx^3}=\frac{d}{dx}\left(\frac{d^2y}{dx^2}\right)=\frac{d}{dx}f''(x)=f'''(x), \text{ etc.}$$

and values of these at $x=a$ are denoted $f''(a)$, $f'''(a)$, etc. (see Table of Derivatives).

2. *Slope of a Curve*

The tangent line at a point $P(x, y)$ of the curve $y=f(x)$ has a slope $f'(x)$ provided that $f'(x)$ exists at P. The slope at P is defined to be that of the tangent line at P. The tangent line at $P(x_1, y_1)$ is given by

$$y - y_1 = f'(x_1)(x - x_1).$$

The *normal line* to the curve at $P(x_1, y_1)$ has slope $-1/f'(x_1)$ and thus obeys the equation

$$y - y_1 = [-1/f'(x_1)](x - x_1)$$

(The slope of a vertical line is not defined.)

3. *Angle of Intersection of Two Curves*

Two curves, $y = f_1(x)$ and $y = f_2(x)$, that intersect at a point $P(X, Y)$ where derivatives $f_1'(X)$, $f_2'(X)$ exist, have an angle (α) of intersection given by

$$\tan \alpha = \frac{f_2'(X) - f_1'(X)}{1 + f_2'(X) \cdot f_1'(X)}.$$

If $\tan \alpha > 0$, then α is the acute angle; if $\tan \alpha < 0$, then α is the obtuse angle.

4. *Radius of Curvature*

The radius of curvature R of the curve $y = f(x)$ at point $P(x, y)$ is

$$R = \frac{\{1 + [f'(x)]^2\}^{3/2}}{f''(x)}$$

In polar coordinates (θ, r) the corresponding formula is

$$R = \frac{\left[r^2 + \left(\frac{dr}{d\theta}\right)^2\right]^{3/2}}{r^2 + 2\left(\frac{dr}{d\theta}\right)^2 - r\frac{d^2r}{d\theta^2}}$$

The *curvature* K is $1/R$.

5. *Relative Maxima and Minima*

The function f has a relative maximum at $x = a$ if $f(a) \geq f(a + c)$ for all values of c (positive or negative) that are sufficiently near zero. The function f has a relative minimum at $x = b$ if $f(b) \leq f(b + c)$ for all values of c that are sufficiently close to zero. If the function f is defined on the closed interval $x_1 \leq x \leq x_2$, and has a relative maximum or minimum at $x = a$, where $x_1 < a < x_2$, and if the derivative $f'(x)$ exists at $x = a$, then $f'(a) = 0$. It is noteworthy that a relative maximum or minimum may occur at a point where the derivative does not exist. Further, the derivative may vanish at a point that is neither a maximum or a minimum for the function. Values of x for which $f'(x) = 0$ are called "critical values." To determine whether a critical value of x, say x_c, is a relative maximum or minimum for the function at x_c, one may use the second derivative test

1. If $f''(x_c)$ is positive, $f(x_c)$ is a minimum

2. If $f''(x_c)$ is negative, $f(x_c)$ is a maximum

3. If $f''(x_c)$ is zero, no conclusion may be made

The sign of the derivative as x advances through x_c may also be used as a test. If $f'(x)$ changes from positive to zero to negative, then a maximum occurs at x_c, whereas a change in $f'(x)$ from negative to zero to positive indicates a minimum. If $f'(x)$ does not change sign as x advances through x_c, then the point is neither a maximum nor a minimum.

6. *Points of Inflection of a Curve*

The sign of the second derivative of f indicates whether the graph of $y=f(x)$ is concave upward or concave downward:

$$f''(x)>0\text{: concave upward}$$

$$f''(x)<0\text{: concave downward}$$

A point of the curve at which the direction of concavity changes is called a point of inflection (Figure 6.1). Such a point may occur where $f''(x)=0$ or where $f''(x)$ becomes infinite. More precisely, if the function $y=f(x)$ and its first derivative $y'=f'(x)$ are continuous in the interval $a \leq x \leq b$, and if $y''=f''(x)$ exists in $a<x<b$, then the graph of $y=f(x)$ for $a<x<b$ is concave

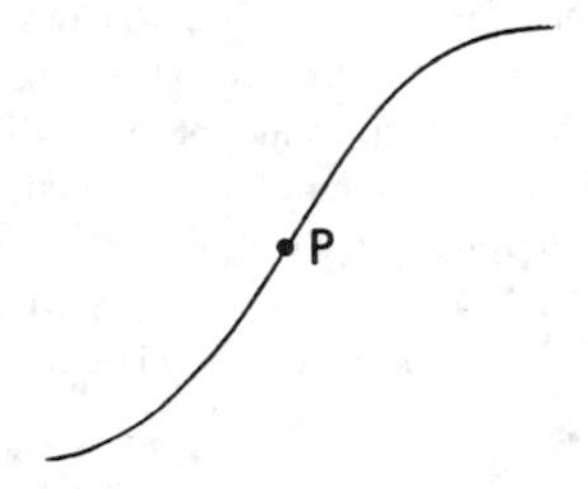

FIGURE 6.1. Point of inflection.

upward if $f''(x)$ is positive and concave downward if $f''(x)$ is negative.

7. *Taylor's Formula*

If f is a function that is continuous on an interval that contains a and x, and if its first $(n+1)$ derivatives are continuous on this interval, then

$$f(x) = f(a) + f'(a)(x-a) + \frac{f''(a)}{2!}(x-a)^2$$

$$+ \frac{f'''(a)}{3!}(x-a)^3 + \dots$$

$$+ \frac{f^{(n)}(a)}{n!}(x-a)^n + R,$$

where R is called the *remainder*. There are various common forms of the remainder:

Lagrange's form:

$$R = f^{(n+1)}(\beta) \cdot \frac{(x-a)^{n+1}}{(n+1)!}; \ \beta \text{ between } a \text{ and } x.$$

Cauchy's form:

$$R = f^{(n+1)}(\beta) \cdot \frac{(x-\beta)^n (x-a)}{n!};$$

β between a and x.

Integral form:

$$R=\int_a^x \frac{(x-t)^n}{n!} f^{(n+1)}(t)\,dt.$$

8. *Indeterminant Forms*

If $f(x)$ and $g(x)$ are continuous in an interval that includes $x=a$ and if $f(a)=0$ and $g(a)=0$, the limit $\lim_{x\to a}(f(x)/g(x))$ takes the form "0/0", called an *indeterminant form*. *L'Hôpital's rule* is

$$\lim_{x\to a}\frac{f(x)}{g(x)}=\lim_{x\to a}\frac{f'(x)}{g'(x)}.$$

Similarly, it may be shown that if $f(x)\to\infty$ and $g(x)\to\infty$ as $x\to a$, then

$$\lim_{x\to a}\frac{f(x)}{g(x)}=\lim_{x\to a}\frac{f'(x)}{g'(x)}.$$

(The above holds for $x\to\infty$.)

Examples

$$\lim_{x\to 0}\frac{\sin x}{x}=\lim_{x\to 0}\frac{\cos x}{1}=1$$

$$\lim_{x\to\infty}\frac{x^2}{e^x}=\lim_{x\to\infty}\frac{2x}{e^x}=\lim_{x\to\infty}\frac{2}{e^x}=0$$

9. *Numerical Methods*

a. *Newton's method* for approximating roots of the equation $f(x)=0$: A first estimate x_1 of the root is made; then provided that $f'(x_1) \neq 0$, a better approximation is x_2

$$x_2 = x_1 - \frac{f(x_1)}{f'(x_1)}.$$

The process may be repeated to yield a third approximation x_3 to the root:

$$x_3 = x_2 - \frac{f(x_2)}{f'(x_2)}.$$

provided $f'(x_2)$ exists. The process may be repeated. (In certain rare cases the process will not converge.)

b. *Trapezoidal rule for areas* (Figure 6.2): For the function $y = f(x)$ defined on the interval (a, b) and positive there, take n equal subintervals of width $\Delta x = (b - a)/n$. The area bounded by the curve between

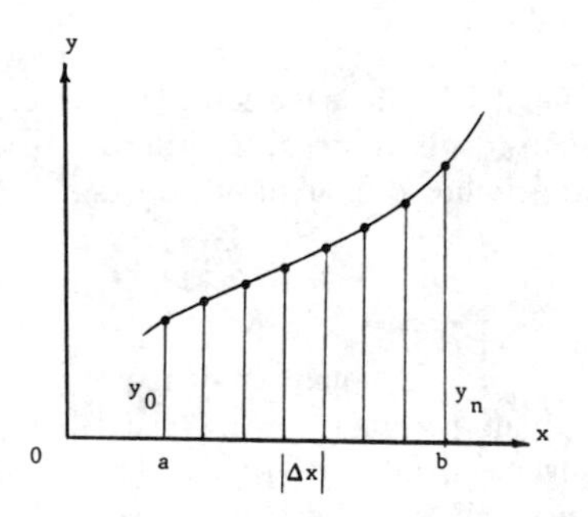

FIGURE 6.2. Trapezoidal rule for area.

$x=a$ and $x=b$ (or definite integral of $f(x)$) is approximately the sum of trapezoidal areas, or

$$A \sim \left(\frac{1}{2}y_0 + y_1 + y_2 + \ldots + y_{n-1} + \frac{1}{2}y_n\right)(\Delta x)$$

Estimation of the error (E) is possible if the second derivative can be obtained:

$$E = \frac{b-a}{12} f''(c)(\Delta x)^2,$$

where c is some number between a and b.

10. *Functions of Two Variables*

For the function of two variables, denoted $z=f(x, y)$, if y is held constant, say at $y=y_1$, then the resulting function is a function of x only. Similarly, x may be held constant at x_1, to give the resulting function of y.

- *The Gas Laws*

A familiar example is afforded by the ideal gas law that relates the pressure p, the volume V and the absolute temperature T of an ideal gas:

$$pV = nRT$$

where n is the number of moles and R is the gas constant per mole, 8.31 ($J \cdot °K^{-1} \cdot mole^{-1}$). By rearrangement, any one of the three variables may be expressed as a function of the other two. Further, either one of these two may be held constant. If T is held constant, then we get the form known as Boyle's law:

$$p = kV^{-1} \qquad \text{(Boyle's law)}$$

where we have denoted nRT by the constant k and, of course, $V > 0$. If the pressure remains constant, we have Charles' law:

$$V = bT \qquad \text{(Charles' law)}$$

where the constant b denotes nR/p. Similarly, volume may be kept constant:

$$p = aT$$

where now the constant, denoted a, is nR/V.

11. Partial Derivatives

The physical example afforded by the ideal gas law permits clear interpretations of processes in which one of the variables is held constant. More generally, we may consider a function $z = f(x, y)$ defined over some region of the x-y-plane in which we hold one of the two coordinates, say y, constant. If the resulting function of x is differentiable at a point (x, y) we denote this derivative by one of the notations

$$f_x, \qquad \delta f/\delta x, \qquad \delta z/\delta x \ .$$

called the *partial derivative with respect to* x. Similarly, if x is held constant and the resulting function of y is differentiable, we get the *partial derivative with respect to* y, denoted by one of the following:

$$f_y \qquad \delta f/\delta y \qquad \delta z/\delta y$$

Example

Given $z = x^4 y^3 - y \sin x + 4y$, then

$$\delta z/\delta x = 4(xy)^3 - y \cos x;$$

$$\delta z/\delta y = 3x^4 y^2 - \sin x + 4.$$

7 Integral Calculus

1. *Indefinite Integral*

If $F(x)$ is differentiable for all values of x in the interval (a, b) and satisfies the equation $dy/dx = f(x)$, then $F(x)$ is an integral of $f(x)$ with respect to x. The notation is $F(x) = \int f(x)\,dx$ or, in differential form, $dF(x) = f(x)\,dx$.

For any function $F(x)$ that is an integral of $f(x)$ it follows that $F(x) + C$ is also an integral. We thus write

$$\int f(x)\,dx = F(x) + C.$$

(See Table of Integrals.)

2. *Definite Integral*

Let $f(x)$ be defined on the interval $[a, b]$ which is partitioned by points $x_1, x_2, \ldots, x_j, \ldots, x_{n-1}$ between $a = x_0$ and $b = x_n$. The jth interval has length $\Delta x_j = x_j - x_{j-1}$, which may vary with j. The sum $\Sigma_{j=1}^{n} f(v_j)\Delta x_j$, where v_j is arbitrarily chosen in the jth subinterval, depends on the numbers $x_0, \ldots, x_n$ and the choice of the v as well as f; but if such sums approach a common value as all Δx approach zero, then this value is the definite integral of f over the interval (a, b) and is denoted $\int_a^b f(x)\,dx$. The *fundamental theorem of integral calculus* states that

$$\int_a^b f(x)\,dx = F(b) - F(a),$$

where F is any continuous indefinite integral of f in the interval (a, b).

3. *Properties*

$$\int_a^b [f_1(x) + f_2(x) + \cdots + f_j(x)]\,dx = \int_a^b f_1(x)\,dx + \int_a^b f_2(x)\,dx + \cdots + \int_a^b f_j(x)\,dx.$$

$$\int_a^b cf(x)\,dx = c\int_a^b f(x)\,dx, \text{ if } c \text{ is a constant.}$$

$$\int_a^b f(x)\,dx = -\int_b^a f(x)\,dx.$$

$$\int_a^b f(x)\,dx = \int_a^c f(x)\,dx + \int_c^b f(x)\,dx.$$

4. *Common Applications of the Definite Integral*

- *Area (Rectangular Coordinates)*

Given the function $y = f(x)$ such that $y > 0$ for all x between a and b, the area bounded by the curve

$y = f(x)$, the x-axis, and the vertical lines $x = a$ and $x = b$ is

$$A = \int_a^b f(x)\, dx.$$

- *Length of Arc (Rectangular Coordinates)*

Given the smooth curve $f(x, y) = 0$ from point (x_1, y_1) to point (x_2, y_2), the length between these points is

$$L = \int_{x_1}^{x_2} \sqrt{1 + (dy/dx)^2}\, dx,$$

$$L = \int_{y_1}^{y_2} \sqrt{1 + (dx/dy)^2}\, dy.$$

- *Mean Value of a Function*

The mean value of a function $f(x)$ continuous on $[a, b]$ is

$$\frac{1}{(b - a)} \int_a^b f(x)\, dx.$$

- *Area (Polar Coordinates)*

Given the curve $r = f(\theta)$, continuous and non-negative for $\theta_1 \le \theta \le \theta_2$, the area enclosed by this curve and the radial lines $\theta = \theta_1$ and $\theta = \theta_2$ is given by

$$A = \int_{\theta_1}^{\theta_2} \frac{1}{2} [f(\theta)]^2\, d\theta.$$

- *Length of Arc (Polar Coordinates)*

Given the curve $r=f(\theta)$ with continuous derivative $f'(\theta)$ on $\theta_1 \leq \theta \leq \theta_2$, the length of arc from $\theta=\theta_1$ to $\theta=\theta_2$ is

$$L=\int_{\theta_1}^{\theta_2}\sqrt{[f(\theta)]^2+[f'(\theta)]^2}\,d\theta.$$

- *Volume of Revolution*

Given a function $y=f(x)$ continuous and non-negative on the interval (a,b), when the region bounded by $f(x)$ between a and b is revolved about the x-axis the volume of revolution is

$$V=\pi\int_a^b[f(x)]^2\,dx.$$

- *Surface Area of Revolution*
  (revolution about the x-axis, between a and b)

If the portion of the curve $y=f(x)$ between $x=a$ and $x=b$ is revolved about the x-axis, the area A of the surface generated is given by the following:

$$A=\int_a^b 2\pi f(x)\{1+[f'(x)]^2\}^{1/2}\,dx$$

- *Work*

If a variable force $f(x)$ is applied to an object in the direction of motion along the x-axis between $x=a$ and $x=b$, the work done is

$$W = \int_a^b f(x)\, dx.$$

5. *Cylindrical and Spherical Coordinates*

a. Cylindrical coordinates (Figure 7.1)

$$x = r \cos \theta$$

$$y = r \sin \theta$$

element of volume $dV = r\,dr\,d\theta\,dz$.

b. Spherical coordinates (Figure 7.2)

$$x = \rho \sin \phi \cos \theta$$

$$y = \rho \sin \phi \sin \theta$$

$$z = \rho \cos \phi$$

element of volume $dV = \rho^2 \sin \phi\, d\rho, d\phi\, d\theta$.

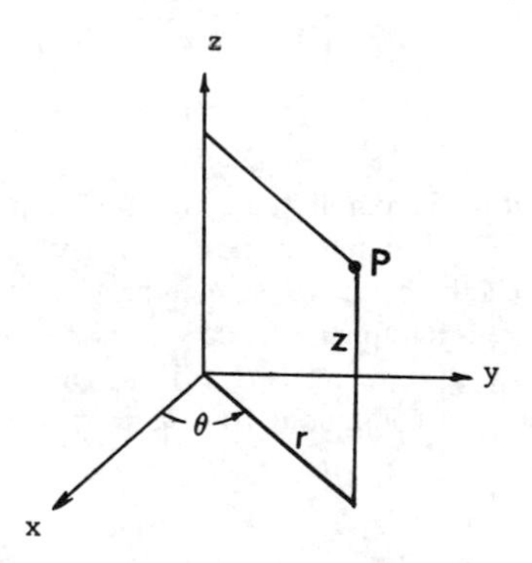

FIGURE 7.1. Cylindrical coordinates.

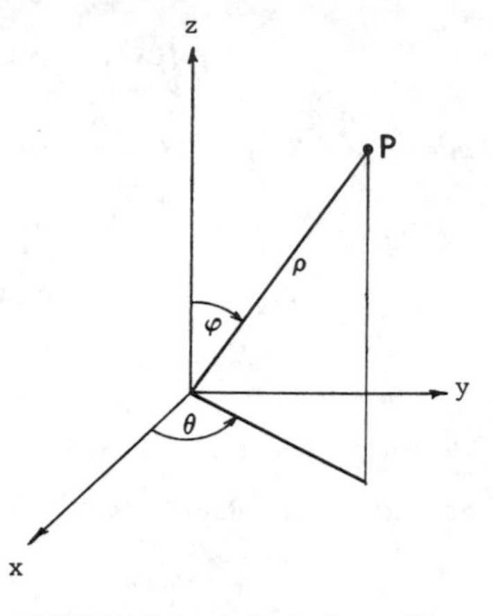

FIGURE 7.2. Spherical coordinates.

6. *Double Integration*

The evaluation of a double integral of $f(x,y)$ over a plane region R

$$\iint_R f(x,y)\,dA$$

is practically accomplished by iterated (repeated) integration. For example, suppose that a vertical straight line meets the boundary of R in at most two points so that there is an upper boundary, $y = y_2(x)$, and a lower boundary, $y = y_1(x)$. Also, it is assumed that these functions are continuous from a to b. (See Fig. 7.3). Then

$$\iint_R f(x,y)\,dA = \int_a^b \left(\int_{y_1(x)}^{y_2(x)} f(x,y)\,dy \right) dx$$

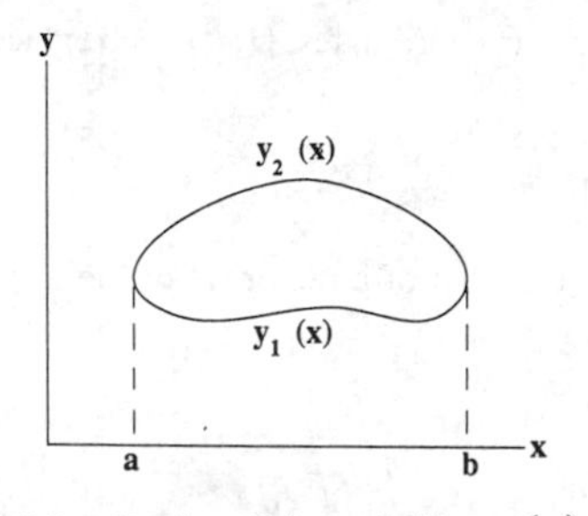

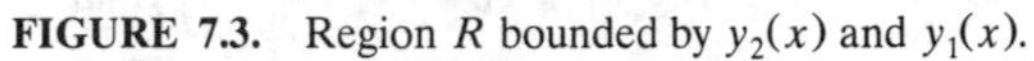

FIGURE 7.3. Region R bounded by $y_2(x)$ and $y_1(x)$.

If R has left-hand boundary, $x = x_1(y)$, and a right-hand boundary, $x = x_2(y)$, which are continuous from c to d (the extreme values of y in R) then

$$\iint_R f(x, y)\, dA = \int_c^d \left(\int_{x_1(y)}^{x_2(y)} f(x, y)\, dx \right) dy$$

Such integrations are sometimes more convenient in polar coordinates, $x = r \cos \theta$, $y = r \sin \theta$; $dA = r\, dr\, d\theta$.

7. *Surface Area and Volume by Double Integration*

For the surface given by $z = f(x, y)$, which projects onto the closed region R of the x–y-plane, one may calculate the volume V bounded above by the surface and below by R, and the surface area S by the following:

$$V = \iint_R z\, dA = \iint_R f(x, y)\, dx\, dy$$

$$S = \iint_R [1 + (\delta z/\delta x)^2 + (\delta z/\delta y)^2]^{1/2}\, dx\, dy$$

[In polar coordinates, (r, θ), we replace dA by $rdrd\theta$].

8. *Centroid*

The centroid of a region R of the x–y-plane is a point (x', y') where

$$x' = \frac{1}{A}\iint_R x\,dA; \qquad y' = \frac{1}{A}\iint_R y\,dA$$

and A is the area of the region.

Example
For the circular sector of angle 2α and radius R, the area A is αR^2; the integral needed for x', expressed in polar coordinates is

$$\iint x\,dA = \int_{-\alpha}^{\alpha}\int_0^R (r\cos\theta)r\,dr\,d\theta$$

$$= \left[\frac{R^3}{3}\sin\theta\right]_{-\alpha}^{+\alpha} = \frac{2}{3}R^3\sin\alpha$$

and thus,

$$x' = \frac{\frac{2}{3}R^3\sin\alpha}{\alpha R^2} = \frac{2}{3}R\frac{\sin\alpha}{\alpha}.$$

Centroids of some common regions are shown below:

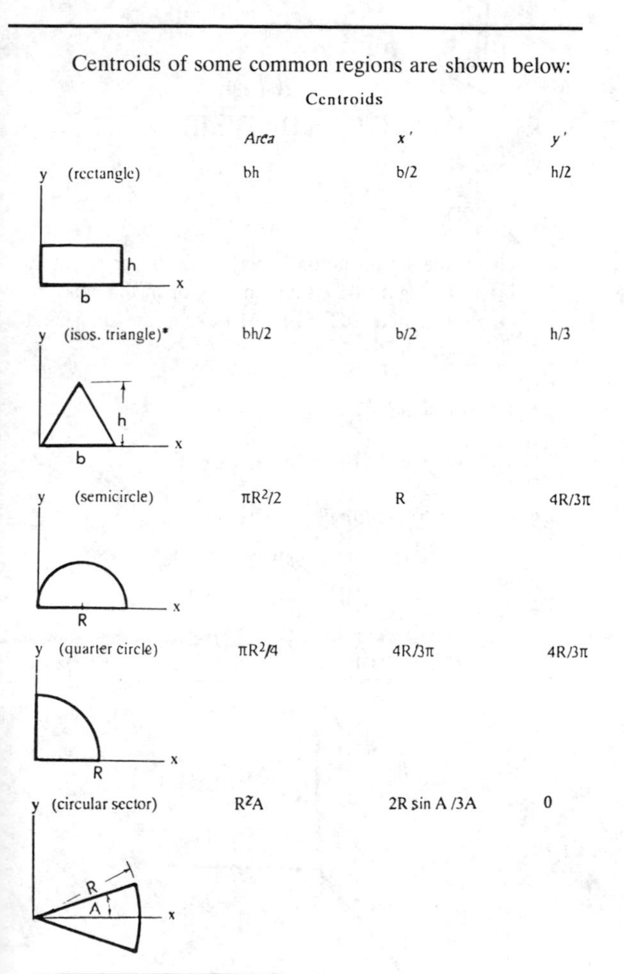

Centroids

	Area	x'	y'
(rectangle)	bh	$b/2$	$h/2$
(isos. triangle)*	$bh/2$	$b/2$	$h/3$
(semicircle)	$\pi R^2/2$	R	$4R/3\pi$
(quarter circle)	$\pi R^2/4$	$4R/3\pi$	$4R/3\pi$
(circular sector)	$R^2 A$	$2R \sin A/3A$	0

* $y' = h/3$ for any triangle of altitude h.

FIGURE 7.4.

8 Vector Analysis

1. *Vectors*

Given the set of mutually perpendicular unit vectors **i**, **j**, and **k** (Figure 8.1), then any vector in the space may be represented as $\mathbf{F} = a\mathbf{i} + b\mathbf{j} + c\mathbf{k}$, where a, b, and c are *components*.

- *Magnitude of* $\mathbf{F}$

$$|\mathbf{F}| = (a^2 + b^2 + c^2)^{\frac{1}{2}}$$

- *Product by scalar* p

$$p\mathbf{F} = pa\mathbf{i} + pb\mathbf{j} + pc\mathbf{k}.$$

- *Sum of* $\mathbf{F}_1$ *and* $\mathbf{F}_2$

$$\mathbf{F}_1 + \mathbf{F}_2 = (a_1 + a_2)\mathbf{i} + (b_1 + b_2)\mathbf{j} + (c_1 + c_2)\mathbf{k}$$

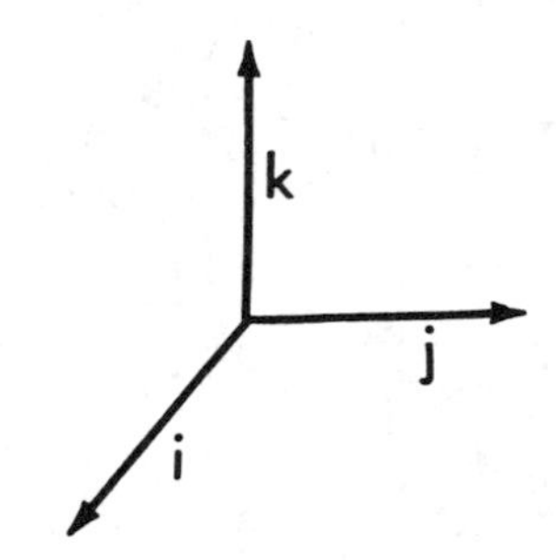

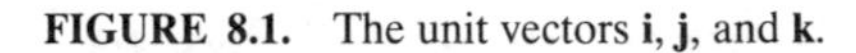
FIGURE 8.1. The unit vectors **i**, **j**, and **k**.

- *Scalar Product*

$$\mathbf{F}_1 \bullet \mathbf{F}_2 = a_1 a_2 + b_1 b_2 + c_1 c_2$$

(Thus, $\mathbf{i} \bullet \mathbf{i} = \mathbf{j} \bullet \mathbf{j} = \mathbf{k} \bullet \mathbf{k} = 1$ and $\mathbf{i} \bullet \mathbf{j} = \mathbf{j} \bullet \mathbf{k} = \mathbf{k} \bullet \mathbf{i} = 0$.)
Also

$$\mathbf{F}_1 \bullet \mathbf{F}_2 = \mathbf{F}_2 \bullet \mathbf{F}_1$$

$$(\mathbf{F}_1 + \mathbf{F}_2) \bullet \mathbf{F}_3 = \mathbf{F}_1 \bullet \mathbf{F}_3 + \mathbf{F}_2 \bullet \mathbf{F}_3$$

- *Vector Product*

$$\mathbf{F}_1 \times \mathbf{F}_2 = \begin{vmatrix} \mathbf{i} & \mathbf{j} & \mathbf{k} \\ a_1 & b_1 & c_1 \\ a_2 & b_2 & c_2 \end{vmatrix}$$

(Thus, $\mathbf{i} \times \mathbf{i} = \mathbf{j} \times \mathbf{j} = \mathbf{k} \times \mathbf{k} = \mathbf{0}$, $\mathbf{i} \times \mathbf{j} = \mathbf{k}$, $\mathbf{j} \times \mathbf{k} = \mathbf{i}$, and $\mathbf{k} \times \mathbf{i} = \mathbf{j}$.)
Also,

$$\mathbf{F}_1 \times \mathbf{F}_2 = -\mathbf{F}_2 \times \mathbf{F}_1$$

$$(\mathbf{F}_1 + \mathbf{F}_2) \times \mathbf{F}_3 = \mathbf{F}_1 \times \mathbf{F}_3 + \mathbf{F}_2 \times \mathbf{F}_3$$

$$\mathbf{F}_1 \times (\mathbf{F}_2 + \mathbf{F}_3) = \mathbf{F}_1 \times \mathbf{F}_2 + \mathbf{F}_1 \times \mathbf{F}_3$$

$$\mathbf{F}_1 \times (\mathbf{F}_2 \times \mathbf{F}_3) = (\mathbf{F}_1 \bullet \mathbf{F}_3)\mathbf{F}_2 - (\mathbf{F}_1 \bullet \mathbf{F}_2)\mathbf{F}_3$$

$$\mathbf{F}_1 \bullet (\mathbf{F}_2 \times \mathbf{F}_3) = (\mathbf{F}_1 \times \mathbf{F}_2) \bullet \mathbf{F}_3$$

2. *Vector Differentiation*

If $\mathbf{V}$ is a vector function of a scalar variable t, then

$$\mathbf{V} = a(t)\mathbf{i} + b(t)\mathbf{j} + c(t)\mathbf{k}$$

and

$$\frac{d\mathbf{V}}{dt} = \frac{da}{dt}\mathbf{i} + \frac{db}{dt}\mathbf{j} + \frac{dc}{dt}\mathbf{k}.$$

For several vector functions $\mathbf{V}_1, \mathbf{V}_2, \ldots, \mathbf{V}_n$

$$\frac{d}{dt}(\mathbf{V}_1 + \mathbf{V}_2 + \ldots + \mathbf{V}_n) = \frac{d\mathbf{V}_1}{dt} + \frac{d\mathbf{V}_2}{dt} + \ldots + \frac{d\mathbf{V}_n}{dt},$$

$$\frac{d}{dt}(\mathbf{V}_1 \bullet \mathbf{V}_2) = \frac{d\mathbf{V}_1}{dt} \bullet \mathbf{V}_2 + \mathbf{V}_1 \bullet \frac{d\mathbf{V}_2}{dt},$$

$$\frac{d}{dt}(\mathbf{V}_1 \times \mathbf{V}_2) = \frac{d\mathbf{V}_1}{dt} \times \mathbf{V}_2 + \mathbf{V}_1 \times \frac{d\mathbf{V}_2}{dt}.$$

For a scalar valued function $g(x, y, z)$

(gradient) $$\text{grad}\, g = \nabla g = \frac{\delta g}{\delta x}\mathbf{i} + \frac{\delta g}{\delta y}\mathbf{j} + \frac{\delta g}{\delta z}\mathbf{k}.$$

For a vector valued function $\mathbf{V}(a, b, c)$, where a, b, c are each a function of x, y, and z,

(divergence) $$\text{div}\,\mathbf{V} = \nabla \bullet \mathbf{V} = \frac{\delta a}{\delta x} + \frac{\delta b}{\delta y} + \frac{\delta c}{\delta z}$$

(curl) $$\text{curl}\,\mathbf{V} = \nabla \times \mathbf{V} = \begin{vmatrix} \mathbf{i} & \mathbf{j} & \mathbf{k} \\ \frac{\delta}{\delta x} & \frac{\delta}{\delta y} & \frac{\delta}{\delta z} \\ a & b & c \end{vmatrix}$$

Also,

$$\operatorname{div}\operatorname{grad} g = \nabla^2 g = \frac{\delta^2 g}{\delta x^2} + \frac{\delta^2 g}{\delta y^2} + \frac{\delta^2 g}{\delta z^2}.$$

and

$$\operatorname{curl}\operatorname{grad} g = \mathbf{0}; \qquad \operatorname{div}\operatorname{curl}\mathbf{V} = 0;$$

$$\operatorname{curl}\operatorname{curl}\mathbf{V} = \operatorname{grad}\operatorname{div}\mathbf{V} - (\mathbf{i}\nabla^2 a + \mathbf{j}\nabla^2 b + \mathbf{k}\nabla^2 c).$$

3. *Divergence Theorem (Gauss)*

Given a vector function F with continuous partial derivatives in a region R bounded by a closed surface S, then

$$\iiint_R \mathit{div}\,\mathbf{F}\,dV = \iint_S \mathbf{n} \bullet \mathbf{F}\,dS,$$

where **n** is the (sectionally continuous) unit normal to S.

4. *Stokes' Theorem*

Given a vector function with continuous gradient over a surface S that consists of portions that are piecewise smooth and bounded by regular closed curves such as C, then

$$\iint_S \mathbf{n} \bullet \operatorname{curl}\mathbf{F}\,dS = \oint_C \mathbf{F} \bullet d\mathbf{r}$$

5. *Planar Motion in Polar Coordinates*

Motion in a plane may be expressed with regard to polar coordinates (r, θ). Denoting the position vector by $\mathbf{r}$ and its magnitude by r, we have $\mathbf{r} = r\mathbf{R}(\theta)$, where $\mathbf{R}$ is the unit vector. Also, $d\mathbf{R}/d\theta = \mathbf{P}$, a unit vector perpendicular to **R**. The velocity and acceleration are then

$$\mathbf{v}=\frac{dr}{dt}\mathbf{R}+r\frac{d\theta}{dt}\mathbf{P};$$

$$\mathbf{a}=\left[\frac{d^2r}{dt^2}-r\left(\frac{d\theta}{dt}\right)^2\right]\mathbf{R}+\left[r\frac{d^2\theta}{dt^2}+2\frac{dr}{dt}\frac{d\theta}{dt}\right]\mathbf{P}.$$

Note that the component of acceleration in the **P** direction (transverse component) may also be written

$$\frac{1}{r}\frac{d}{dt}\left(r^2\frac{d\theta}{dt}\right)$$

so that in purely radial motion it is zero and

$$r^2\frac{d\theta}{dt}=C \text{ (constant)}$$

which means that the position vector sweeps out area at a constant rate (see Area in Polar Coordinates, Section 7.4).

9 Special Functions

1. *Hyperbolic Functions*

$$\sinh x = \frac{e^x - e^{-x}}{2} \qquad \operatorname{csch} x = \frac{1}{\sinh x}$$

$$\cosh x = \frac{e^x + e^{-x}}{2} \qquad \operatorname{sech} x = \frac{1}{\cosh x}$$

$$\tanh x = \frac{e^x - e^{-x}}{e^x + e^{-x}} \qquad \operatorname{ctnh} x = \frac{1}{\tanh x}$$

$$\sinh(-x) = -\sinh x \qquad \operatorname{ctnh}(-x) = -\operatorname{ctnh} x$$

$$\cosh(-x) = \cosh x \qquad \operatorname{sech}(-x) = \operatorname{sech} x$$

$$\tanh(-x) = -\tanh x \qquad \operatorname{csch}(-x) = -\operatorname{csch} x$$

$$\tanh x = \frac{\sinh x}{\cosh x} \qquad \operatorname{ctnh} x = \frac{\cosh x}{\sinh x}$$

$$\cosh^2 x - \sinh^2 x = 1 \qquad \cosh^2 x = \frac{1}{2}(\cosh 2x + 1)$$

$$\sinh^2 x = \frac{1}{2}(\cosh 2x - 1) \qquad \operatorname{ctnh}^2 x - \operatorname{csch}^2 x = 1$$

$$\operatorname{csch}^2 x - \operatorname{sech}^2 x = \operatorname{csch}^2 x \operatorname{sech}^2 x \qquad \tanh^2 x + \operatorname{sech}^2 x = 1$$

$$\sinh(x+y) = \sinh x \cosh y + \cosh x \sinh y$$

$$\cosh(x+y) = \cosh x \cosh y + \sinh x \sinh y$$

$$\sinh(x-y) = \sinh x \cosh y - \cosh x \sinh y$$

$$\cosh(x-y) = \cosh x \cosh y - \sinh x \sinh y$$

$$\tanh(x+y) = \frac{\tanh x + \tanh y}{1 + \tanh x \tanh y}$$

$$\tanh(x-y) = \frac{\tanh x - \tanh y}{1 - \tanh x \tanh y}$$

2. *Gamma Function (Generalized Factorial Function)*

The gamma function, denoted $\Gamma(x)$, is defined by

$$\Gamma(x) = \int_0^\infty e^{-t} t^{x-1}\, dt, \qquad x > 0$$

- *Properties*

$$\Gamma(x+1) = x\Gamma(x), \qquad x > 0$$

$$\Gamma(1) = 1$$

$$\Gamma(n+1) = n\Gamma(n) = n!, \qquad (n = 1, 2, 3, \ldots)$$

$$\Gamma(x)\Gamma(1-x) = \pi/\sin \pi x$$

$$\Gamma\left(\frac{1}{2}\right) = \sqrt{\pi}$$

$$2^{2x-1}\Gamma(x)\Gamma\left(x + \frac{1}{2}\right) = \sqrt{\pi}\,\Gamma(2x)$$

3. *Laplace Transforms*

The Laplace transform of the function $f(t)$, denoted by $F(s)$ or $L\{f(t)\}$, is defined

$$F(s)=\int_0^{\infty} f(t)e^{-st}\,dt$$

provided that the integration may be validly performed. A sufficient condition for the existence of $F(s)$ is that $f(t)$ be of exponential order as $t\to\infty$ and that it is sectionally continuous over every finite interval in the range $t\geq 0$. The Laplace transform of $g(t)$ is denoted by $L\{g(t)\}$ or $G(s)$.

- *Operations*

$f(t)$	$F(s)=\int_0^{\infty} f(t)e^{-st}\,dt$
$af(t)+bg(t)$	$aF(s)+bG(s)$
$f'(t)$	$sF(s)-f(0)$
$f''(t)$	$s^2F(s)-sf(0)-f'(0)$
$f^{(n)}(t)$	$s^nF(s)-s^{n-1}f(0)$ $-s^{n-2}f'(0)$ $-\cdots-f^{(n-1)}(0)$
$tf(t)$	$-F'(s)$
$t^nf(t)$	$(-1)^nF^{(n)}(s)$
$e^{at}f(t)$	$F(s-a)$

$\int_0^t f(t-\beta)\cdot g(\beta)\,d\beta$	$F(s)\cdot G(s)$
$f(t-a)$	$e^{-as}F(s)$
$f\left(\frac{t}{a}\right)$	$aF(as)$
$\int_0^t g(\beta)\,d\beta$	$\frac{1}{s}G(s)$
$f(t-c)\delta(t-c)$	$e^{-cs}F(s), c>0$

where

$$\delta(t-c)=0 \text{ if } 0\le t<c$$
$$=1 \text{ if } t\ge c$$

$f(t)=f(t+\omega)$ (periodic)	$\dfrac{\int_0^\omega e^{-s\tau}f(\tau)\,d\tau}{1-e^{-s\omega}}$

- *Table of Laplace Transforms*

$f(t)$	$F(s)$	
1	$1/s$	
t	$1/s^2$	
$\dfrac{t^{n-1}}{(n-1)!}$	$1/s^n$	$(n=1,2,3,\ldots)$
$\sqrt{t}$	$\dfrac{1}{2s}\sqrt{\dfrac{\pi}{s}}$	

$\frac{1}{\sqrt{t}}$	$\sqrt{\frac{\pi}{s}}$	
e^{at}	$\frac{1}{s-a}$	
te^{at}	$\frac{1}{(s-a)^2}$	
$\frac{t^{n-1}e^{at}}{(n-1)!}$	$\frac{1}{(s-a)^n}$	$(n=1,2,3,\ldots)$
$\frac{t^x}{\Gamma(x+1)}$	$\frac{1}{s^{x+1}}$,	$x>-1$
$\sin at$	$\frac{a}{s^2+a^2}$	
$\cos at$	$\frac{s}{s^2+a^2}$	
$\sinh at$	$\frac{a}{s^2-a^2}$	
$\cosh at$	$\frac{s}{s^2-a^2}$	
$e^{at}-e^{bt}$	$\frac{a-b}{(s-a)(s-b)}$,	$(a\neq b)$
$ae^{at}-be^{bt}$	$\frac{s(a-b)}{(s-a)(s-b)}$,	$(a\neq b)$
$t\sin at$	$\frac{2as}{(s^2+a^2)^2}$	
$t\cos at$	$\frac{s^2-a^2}{(s^2+a^2)^2}$	

$e^{at} \sin bt$	$\dfrac{b}{(s-a)^2+b^2}$
$e^{at} \cos bt$	$\dfrac{s-a}{(s-a)^2+b^2}$
$\dfrac{\sin at}{t}$	$Arc \tan \dfrac{a}{s}$
$\dfrac{\sinh at}{t}$	$\dfrac{1}{2}\log_e\left(\dfrac{s+a}{s-a}\right)$

4. *Z-Transform*

For the real-valued sequence $\{f(k)\}$ and complex variable z, the z-transform, $F(z) = Z\{f(k)\}$ is defined by

$$Z\{f(k)\} = F(z) = \sum_{k=0}^{\infty} f(k) z^{-k}$$

For example, the sequence $f(k) = 1$, $k = 0, 1, 2, \ldots$, has the z-transform

$$F(z) = 1 + z^{-1} + z^{-2} + z^{-3} \ldots + z^{-k} + \ldots.$$

- ***z-Transform and the Laplace Transform***

For function $U(t)$ the output of the ideal sampler $U^*(t)$ is a set of values $U(kT)$, $k = 0, 1, 2, \ldots$, that is,

$$U^*(t) = \sum_{k=0}^{\infty} U(t)\, \delta(t - kT)$$

The Laplace transform of the output is

$$\mathcal{L}\{U^*(t)\} = \int_0^\infty e^{-st} U^*(t)\, dt = \int_0^\infty e^{-st} \sum_{k=0}^{\infty} U(t)\delta(t-kT)\, dt$$

$$= \sum_{k=0}^{\infty} e^{-skT}\, U(kT)$$

Defining $z = e^{sT}$ gives

$$\mathcal{L}\{U^*(t)\} = \sum_{k=0}^{\infty} U(kT) z^{-k}$$

which is the z-transform of the sampled signal $U(kT)$.

- *Properties*

Linearity: $Z\{af_1(k) + bf_2(k)\} = aZ\{f_1(k)\} + bZ\{f_2(k)\}$
$= aF_1(z) + bF_2(z)$

Right-shifting property: $Z\{f(k-n)\} = z^{-n}F(z)$

Left-shifting property: $Z\{f(k+n)\} = z^n F(z) - \sum_{k=0}^{n-1} f(k) z^{n-k}$

Time scaling: $Z\{a^k f(k)\} = F(z/a)$

Multiplication by k: $Z\{kf(k)\} = -z\,dF(z)/dz$

Initial value: $f(0) = \lim_{z\to\infty} (1 - z^{-1})F(z) = F(\infty)$

Final value: $\lim_{k\to\infty} f(k) = \lim_{z\to 1} (1 - z^{-1})F(z)$

Convolution: $Z\{f_1(k)^* f_2(k)\} = F_1(z)F_2(z)$

- *z-Transforms of Sampled Functions*

$f(k)$	$Z\{f(kT)\} = F(z)$
1 at k; else 0	z^{-k}
1	$\dfrac{z}{z-1}$
kT	$\dfrac{Tz}{(z-1)^2}$
$(kT)^2$	$\dfrac{T^2 z(z+1)}{(z-1)^3}$
$\sin \omega kT$	$\dfrac{z \sin \omega T}{z^2 - 2z\cos \omega T + 1}$
$\cos \omega T$	$\dfrac{z(z - \cos \omega T)}{z^2 - 2z\cos \omega T + 1}$
e^{-akT}	$\dfrac{z}{z - e^{-aT}}$
kTe^{-akT}	$\dfrac{zTe^{-aT}}{(z - e^{-aT})^2}$

$(kT)^2 e^{-akT}$	$\dfrac{T^2 e^{-aT} z(z+e^{-aT})}{(z-e^{-aT})^3}$
$e^{-akT} \sin \omega kT$	$\dfrac{ze^{-aT} \sin \omega T}{z^2 - 2ze^{-aT} \cos \omega T + e^{-2aT}}$
$e^{-akT} \cos \omega kT$	$\dfrac{z(z-e^{-aT} \cos \omega T)}{z^2 - 2ze^{-aT} \cos \omega T + e^{-2aT}}$
$a^k \sin \omega kT$	$\dfrac{az \sin \omega T}{z^2 - 2az \cos \omega T + a^2}$
$a^k \cos \omega kT$	$\dfrac{z(z - a \cos \omega T)}{z^2 - 2az \cos \omega T + a^2}$

5. *Fourier Series*

The periodic function $f(t)$, with period 2π may be represented by the trigonometric series

$$a_0 + \sum_1^{\infty} (a_n \cos nt + b_n \sin nt)$$

where the coefficients are determined from

$$a_0 = \frac{1}{2\pi} \int_{-\pi}^{\pi} f(t)\, dt$$

$$a_n = \frac{1}{\pi} \int_{-\pi}^{\pi} f(t) \cos nt\, dt$$

$$b_n = \frac{1}{\pi} \int_{-\pi}^{\pi} f(t) \sin nt\, dt \qquad (n = 1, 2, 3, \dots)$$

Such a trigonometric series is called the Fourier series corresponding to $f(t)$ and the coefficients are termed Fourier coefficients of $f(t)$. If the function is piecewise continuous in the interval $-\pi \leq t \leq \pi$, and has left- and right-hand derivatives at each point in that interval, then the series is convergent with sum $f(t)$ except at points t_i at which $f(t)$ is discontinuous. At such points of discontinuity, the sum of the series is the arithmetic mean of the right- and left-hand limits of $f(t)$ at t_i. The integrals in the formulas for the Fourier coefficients can have limits of integration that span a length of 2π, for example, 0 to 2π (because of the periodicity of the integrands).

6. *Functions with Period Other Than 2π*

If $f(t)$ has period P the Fourier series is

$$f(t) \sim a_0 + \sum_1^{\infty} \left(a_n \cos \frac{2\pi n}{P} t + b_n \sin \frac{2\pi n}{P} t \right),$$

where

$$a_0 = \frac{1}{P} \int_{-P/2}^{P/2} f(t)\, dt$$

$$a_n = \frac{2}{P} \int_{-P/2}^{P/2} f(t) \cos \frac{2\pi n}{P} t\, dt$$

$$b_n = \frac{2}{P} \int_{-P/2}^{P/2} f(t) \sin \frac{2\pi n}{P} t\, dt.$$

Again, the interval of integration in these formulas may be replaced by an interval of length P, for example, 0 to P.

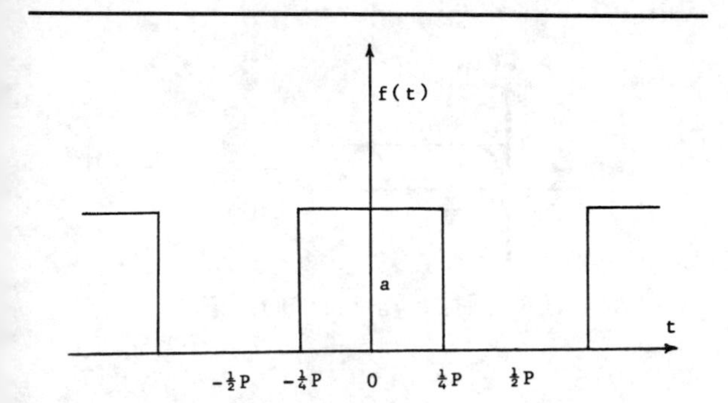

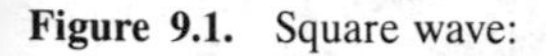

Figure 9.1. Square wave:

$$f(t) \sim \frac{a}{2} + \frac{2a}{\pi}\left(\cos\frac{2\pi t}{P} - \tfrac{1}{3}\cos\frac{6\pi t}{P} + \tfrac{1}{5}\cos\frac{10\pi t}{P} + \ldots\right).$$

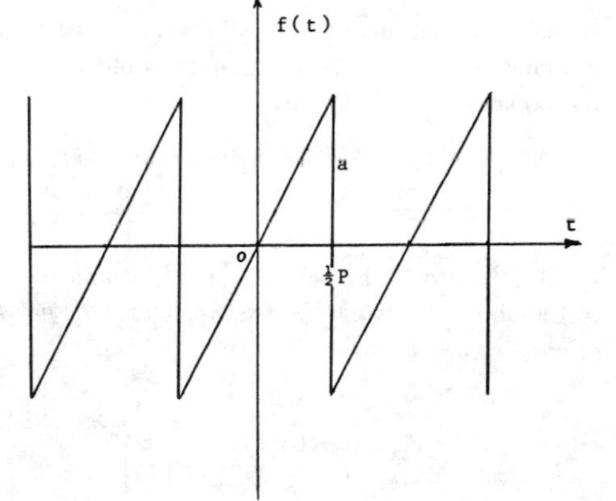

FIGURE 9.2. Sawtooth wave:

$$f(t) \sim \frac{2a}{\pi}\left(\sin\frac{2\pi t}{P} - \tfrac{1}{2}\sin\frac{4\pi t}{P} + \tfrac{1}{3}\sin\frac{6\pi t}{P} - \ldots\right).$$

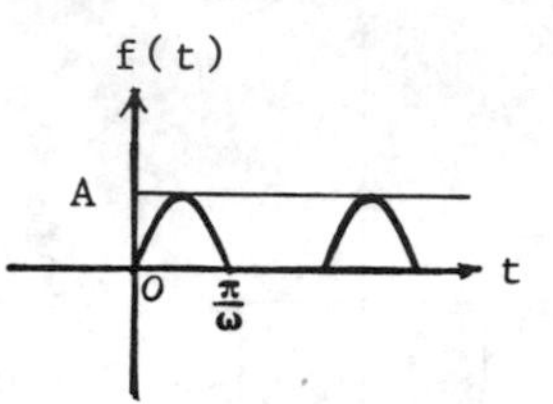

FIGURE 9.3. Half-wave rectifier:

$$f(t) \sim \frac{A}{\pi} + \frac{A}{2}\sin \omega t -$$

$$\frac{2A}{\pi}\left(\frac{1}{(1)(3)}\cos 2\omega t + \frac{1}{(3)(5)}\cos 4\omega t + \dots\right).$$

7. *Bessel Functions*

Bessel functions, also called cylindrical functions, arise in many physical problems as solutions of the differential equation

$$x^2y'' + xy' + (x^2 - n^2)y = 0$$

which is known as Bessel's equation. Certain solutions of the above, known as *Bessel functions of the first kind of order n*, are given by

$$J_n(x) = \sum_{k=0}^{\infty} \frac{(-1)^k}{k!\Gamma(n+k+1)}\left(\frac{x}{2}\right)^{n+2k}$$

$$J_{-n}(x) = \sum_{k=0}^{\infty} \frac{(-1)^k}{k!\Gamma(-n+k+1)}\left(\frac{x}{2}\right)^{-n+2k}$$

In the above it is noteworthy that the gamma function must be defined for the negative argument q: $\Gamma(q) = \Gamma(q+1)/q$, provided that q is not a negative integer. When q is a negative integer, $1/\Gamma(q)$ is defined to be zero. The functions $J_{-n}(x)$ and $J_n(x)$ are solutions of Bessel's equation for all real n. It is seen, for $n = 1, 2, 3, \ldots$ that

$$J_{-n}(x) = (-1)^n J_n(x)$$

and, therefore, these are not independent; hence, a linear combination of these is not a general solution. When, however, n is not a positive integer, a negative integer, nor zero, the linear combination with arbitrary constants c_1 and c_2

$$y = c_1 J_n(x) + c_2 J_{-n}(x)$$

is the general solution of the Bessel differential equation.

The zero order function is especially important as it arises in the solution of the heat equation (for a "long" cylinder):

$$J_0(x) = 1 - \frac{x^2}{2^2} + \frac{x^4}{2^2 4^2} - \frac{x^6}{2^2 4^2 6^2} + \ldots$$

while the following relations show a connection to the trigonometric functions:

$$J_{\frac{1}{2}}(x) = \left[\frac{2}{\pi x}\right]^{1/2} \sin x$$

$$J_{-\frac{1}{2}}(x)=\left[\frac{2}{\pi x}\right]^{1/2} \cos x$$

The following recursion formula gives $J_{n+1}(x)$ for any order in terms of lower order functions:

$$\frac{2n}{x}J_n(x)=J_{n-1}(x)+J_{n+1}(x)$$

8. *Legendre Polynomials*

If Laplace's equation, $\nabla^2 V=0$, is expressed in spherical coordinates, it is

$$r^2 \sin\theta\frac{\delta^2 V}{\delta r^2}+2r\sin\theta\frac{\delta V}{\delta r}+\sin\theta\frac{\delta^2 V}{\delta\theta^2}+\cos\theta\frac{\delta V}{\delta\theta}$$

$$+\frac{1}{\sin\theta}\frac{\delta^2 V}{\delta\phi^2}=0$$

and any of its solutions, $V(r, \theta, \phi)$, are known as *spherical harmonics*. The solution as a product

$$V(r,\theta,\phi)=R(r)\Theta(\theta)$$

which is independent of ϕ, leads to

$$\sin^2\theta\Theta''+\sin\theta\cos\theta\Theta'+[n(n+1)\sin^2\theta]\Theta=0$$

Rearrangement and substitution of $x=\cos\theta$ leads to

$$(1-x^2)\frac{d^2\Theta}{dx^2}-2x\frac{d\Theta}{dx}+n(n+1)\Theta=0$$

known as *Legendre's equation.* Important special cases are those in which n is zero or a positive integer, and for such cases, Legendre's equation is satisfied by poly nomials called Legendre polynomials, $P_n(x)$. A short list of Legendre polynomials, expressed in terms of x and $\cos\theta$, is given below. These are given by the following general formula:

$$P_n(x) = \sum_{j=0}^{L} \frac{(-1)^j (2n-2j)!}{2^n j!(n-j)!(n-2j)!} x^{n-2j}$$

where $L = n/2$ if n is even and $L = (n-1)/2$ if n is odd. Some are given below:

$$P_0(x) = 1$$

$$P_1(x) = x$$

$$P_2(x) = \frac{1}{2}(3x^2 - 1)$$

$$P_3(x) = \frac{1}{2}(5x^3 - 3x)$$

$$P_4(x) = \frac{1}{8}(35x^4 - 30x^2 + 3)$$

$$P_5(x) = \frac{1}{8}(63x^5 - 70x^3 + 15x)$$

$$P_0(\cos\theta) = 1$$

$$P_1(\cos\theta) = \cos\theta$$

$$P_2(\cos\theta) = \frac{1}{4}(3\cos 2\theta + 1)$$

$$P_3(\cos\theta) = \frac{1}{8}(5\cos 3\theta + 3\cos\theta)$$

$$P_4(\cos\theta) = \frac{1}{64}(35\cos 4\theta + 20\cos 2\theta + 9)$$

Additional Legendre polynomials may be determined from the *recursion formula*

$$(n+1)P_{n+1}(x) - (2n+1)xP_n(x) + nP_{n-1}(x) = 0 \qquad (n = 1, 2, \ldots)$$

or the *Rodrigues formula*

$$P_n(x) = \frac{1}{2^n n!}\frac{d^n}{dx^n}(x^2-1)^n$$

9. *Laguerre Polynomials*

Laguerre polynomials, denoted $L_n(x)$, are solutions of the differential equation

$$xy'' + (1-x)y' + ny = 0$$

and are given by

$$L_n(x) = \sum_{j=0}^{n} \frac{(-1)^j}{j!} C_{(n,j)} x^j \qquad (n = 0, 1, 2, \ldots)$$

Thus,

$$L_0(x) = 1$$

$$L_1(x) = 1 - x$$

$$L_2(x) = 1 - 2x + \frac{1}{2}x^2$$

$$L_3(x) = 1 - 3x + \frac{3}{2}x^2 - \frac{1}{6}x^3$$

Additional Laguerre polynomials may be obtained from the recursion formula

$$(n+1)L_{n+1}(x)-(2n+1-x)L_n(x)$$
$$+nL_{n-1}(x)=0$$

10. Hermite Polynomials

The Hermite polynomials, denoted $H_n(x)$, are given by

$$H_0=1, \quad H_n(x)=(-1)^n\, e^{x^2}\frac{d^n e^{-x^2}}{dx^n},$$
$$(n=1,2,\ldots)$$

and are solutions of the differential equation

$$y''-2xy'+2ny=0 \qquad (n=0,1,2,\ldots)$$

The first few Hermite polynomials are

$$H_0=1 \qquad H_1(x)=2x$$
$$H_2(x)=4x^2-2 \qquad H_3(x)=8x^3-12x$$
$$H_4(x)=16x^4-48x^2+12$$

Additional Hermite polynomials may be obtained from the relation

$$H_{n+1}(x)=2xH_n(x)-H_n'(x),$$

where prime denotes differentiation with respect to x.

11. *Orthogonality*

A set of functions $\{f_n(x)\}$ $(n=1,2,\ldots)$ is orthogonal in an interval (a,b) with respect to a given weight function $w(x)$ if

$$\int_a^b w(x) f_m(x) f_n(x)\,dx = 0 \qquad \text{when } m \neq n$$

The following polynomials are orthogonal on the given interval for the given $w(x)$:

Legendre polynomials: $P_n(x)$ $\quad w(x)=1$
$a=-1,\ b=1$

Laguerre polynomials: $L_n(x)$ $\quad w(x)=\exp(-x)$
$a=0,\ b=\infty$

Hermite polynomials: $H_n(x)$ $\quad w(x)=\exp(-x^2)$
$a=-\infty,\ b=\infty$

The Bessel functions *of order n*, $J_n(\lambda_1 x), J_n(\lambda_2 x), \ldots,$ are orthogonal with respect to $w(x)=x$ over the interval $(0,c)$ provided that the λ_i are the positive roots of $J_n(\lambda c)=0$:

$$\int_0^c x J_n(\lambda_j x) J_n(\lambda_k x)\,dx = 0 \qquad (j \neq k)$$

where n is fixed and $n \geq 0$.

10 Differential Equations

1. First Order-First Degree Equations

$$M(x, y)\,dx + N(x, y)\,dy = 0$$

a. If the equation can be put in the form $A(x)\,dx + B(y)\,dy = 0$, it is *separable* and the solution follows by integration: $\int A(x)\,dx + \int B(y)\,dy = C$; thus, $x(1+y^2)\,dx + y\,dy = 0$ is separable since it is equivalent to $x\,dx + y\,dy/(1+y^2) = 0$, and integration yields $x^2/2 + \frac{1}{2}\log(1+y^2) + C = 0$.

b. If $M(x, y)$ and $N(x, y)$ are *homogeneous* and of the *same degree* in x and y, then substitution of vx for y (thus, $dy = v\,dx + x\,dv$) will yield a separable equation in the variables x and y. [A function such as $M(x, y)$ is homogeneous of degree n in x and y if $M(cx, cy) = c^n M(x, y)$.] For example, $(y-2x)dx + (2y+x)dy$ has M and N each homogeneous and of degree one so that substitution of $y = vx$ yields the separable equation

$$\frac{2}{x}\,dx + \frac{2v+1}{v^2+v-1}\,dv = 0.$$

c. If $M(x, y)\,dx + N(x, y)\,dy$ is the differential of some function $F(x, y)$, then the given equation is said to be *exact*. A necessary and sufficient condition for exactness is $\partial M/\partial y = \partial N/\partial x$. When the equation is exact, F is found from the relations $\partial F/\partial x = M$ and $\partial F/\partial y = N$, and the solution is $F(x, y) = C$ (constant). For example, $(x^2 + y)\,dy + (2xy - 3x^2)\,dx$ is exact since $\partial M/\partial y = 2x$ and $\partial N/\partial x = 2x$. F is found from $\partial F/\partial x = 2xy - 3x^2$ and $\partial F/\partial y = x^2 + y$. From the first of these, $F = x^2y - x^3 + \phi(y)$; from the second, $F = x^2y + y^2/2 + \Psi(x)$. It follows that $F = x^2y - x^3 + y^2/2$, and $F = C$ is the solution.

d. Linear, order one in y: Such an equation has the form $dy + P(x)y\,dx = Q(x)\,dx$. Multiplication by $\exp[\int P(x)\,dx]$ yields

$$d\left[y\exp\left(\int P\,dx\right)\right] = Q(x)\exp\left(\int P\,dx\right)dx.$$

For example, $dy + (2/x)y\,dy = x^2\,dx$ is linear in y. $P(x) = 2/x$, so $\int P\,dx = 2\ln x = \ln x^2$, and $\exp(\int P\,dx) = x^2$. Multiplication by x^2 yields $d(x^2y) = x^4\,dx$, and integration gives the solution $x^2y = x^5/5 + C$.

2. *Second Order Linear Equations (With Constant Coefficients)*

$$(b_0D^2 + b_1D + b_2)y = f(x), \qquad D = \frac{d}{dx}.$$

a. Right-hand side $= 0$ (homogeneous case)

$$(b_0D^2 + b_1D + b_2)y = 0.$$

The *auxiliary equation* associated with the above is

$$b_0m^2+b_1m+b_2=0.$$

If the roots of the auxiliary equation are *real and distinct*, say m_1 and m_2, then the solution is

$$y=C_1e^{m_1x}+C_2e^{m_2x}$$

where the C's are arbitrary constants.

If the roots of the auxiliary equation are *real and repeated*, say $m_1=m_2=p$, then the solution is

$$y=C_1e^{px}+C_2xe^{px}.$$

If the roots of the auxiliary equation are *complex* $a+ib$ and $a-ib$, then the solution is

$$y=C_1e^{ax}\cos bx+C_2e^{ax}\sin bx.$$

b. Right-hand side $\neq 0$ (nonhomogeneous case)

$$(b_0D^2+b_1D+b_2)y=f(x)$$

The general solution is $y=C_1y_1(x)+C_2y_2(x)+y_p(x)$ where y_1 and y_2 are solutions of the corresponding homogeneous equation and y_p is a solution of the given nonhomogeneous differential equation. y_p has the form $y_p(x)=A(x)y_1(x)+B(x)y_2(x)$ and A and B are found from simultaneous solution of $A'y_1+B'y_2=0$ and $A'y_1'+B'y_2'=f(x)/b_0$. A solution exists if the determinant

$$\begin{vmatrix} y_1 & y_2 \\ y_1' & y_2' \end{vmatrix}$$

does not equal zero. The simultaneous equations yield A' and B' from which A and B follow by integration. For example,

$$(D^2+D-2)y=e^{-3x}.$$

The auxiliary equation has the distinct roots 1 and -2; hence $y_1=e^x$ and $y_2=e^{-2x}$, so that $y_p=Ae^x+Be^{-2x}$. The simultaneous equations are

$$A'e^x-2B'e^{-2x}=e^{-3x}$$

$$A'e^x+B'e^{-2x}=0$$

and give $A'=(1/3)e^{-4x}$ and $B'=(-1/3)e^{-x}$. Thus, $A=(-1/12)e^{-4x}$ and $B=(1/3)e^{-x}$ so that

$$y_p=(-1/12)e^{-3x}+(1/3)e^{-3x}$$

$$=\tfrac{1}{4}e^{-3x}.$$

$$\therefore y=C_1e^x+C_2e^{-2x}+\tfrac{1}{4}e^{-3x}.$$

11 Statistics

1. *Arithmetic Mean*

$$\mu = \frac{\Sigma X_i}{N},$$

where X_i is a measurement in the population and N is the total number of X_i in the population. For a *sample* of size n the sample mean, denoted $\bar{X}$, is

$$\bar{X} = \frac{\Sigma X_i}{n}.$$

2. *Median*

The median is the middle measurement when an odd number (n) measurements is arranged in order; if n is even, it is the midpoint between the two middle measurements.

3. *Mode*

It is the most frequently occurring measurement in a set.

4. *Geometric Mean*

$$\text{geometric mean} = \sqrt[n]{X_1 X_2 \ldots X_n}$$

5. *Harmonic Mean*

The Harmonic mean H of n numbers $X_1, X_2, \ldots, X_n$, is

$$H = \frac{n}{\Sigma(1/Xi)}$$

6. *Variance*

The mean of the sum of squares of deviations from the mean (μ) is the population variance, denoted σ^2

$$\sigma^2 = \Sigma(X_i - \mu)^2/N.$$

The sample variance, s^2, for sample size n is

$$s^2 = \Sigma(X_i - \overline{X})^2/(n-1).$$

A simpler computational form is

$$s^2 = \frac{\Sigma X_i^2 - \frac{(\Sigma X_i)^2}{n}}{n-1}$$

7. *Standard Deviation*

The positive square root of the population variance is the standard deviation. For a population

$$\sigma = \left[\frac{\Sigma X_i^2 - \frac{(\Sigma X_i)^2}{N}}{N}\right]^{1/2};$$

for a sample

$$s = \left[\frac{\Sigma X_i^2 - \frac{(\Sigma X_i)^2}{n}}{n-1} \right]^{1/2}.$$

8. *Coefficient of Variation*

$$V = s/\bar{X}.$$

9. *Probability*

For the sample space U, with subsets A of U (called "events"), we consider the probability measure of an event A to be a real-valued function p defined over all subsets of U such that:

$0 \leq p(A) \leq 1$
$p(U) = 1$ and $p(\Phi) = 0$
If A_1 and A_2 are subsets of U
$p(A_1 \cup A_2) = p(A_1) + p(A_2) - p(A_1 \cap A_2)$

Two events A_1 and A_2 are called mutually exclusive if and only if $A_1 \cap A_2 = \phi$ (null set). These events are said to be independent if and only if $p(A_1 \cap A_2) = p(A_1)p(A_2)$.

- *Conditional Probability and Bayes' Rule*

The probability of an event A, given that an event B has occurred, is called the conditional probability and is denoted $p(A/B)$. Further

$$p(A/B) = \frac{p(A \cap B)}{p(B)}$$

Bayes' rule permits a calculation of *a posteriori* probability from given *a priori* probabilities and is stated below:

If $A_1, A_2, \ldots, A_n$ are n mutually exclusive events, and $p(A_1)+p(A_2)+ \ldots +p(A_n)=1$, and B is any event such that $p(B)$ is not 0, then the conditional probability $p(A_i/B)$ for any one of the events A_i, *given that B has occurred* is

$$p(A_i/B)=\frac{p(A_i)p(B/A_i)}{p(A_1)p(B/A_1)+p(A_2)p(B/A_2)+\ldots+p(A_n)p(B/A_n)}$$

Example

Among 5 different laboratory tests for detecting a certain disease, one is effective with probability 0.75, whereas each of the others is effective with probability 0.40. A medical student, unfamiliar with the advantage of the best test, selects one of them and is successful in detecting the disease in a patient. What is the probability that the most effective test was used?

Let B denote (the event) of detecting the disease, A_1 the selection of the best test, and A_2 the selection of one of the other 4 tests; thus, $p(A_1)=1/5$, $p(A_2)=4/5$, $p(B/A_1)=0.75$ and $p(B/A_2)=0.40$. Therefore

$$p(A_1/B)=\frac{\frac{1}{5}(0.75)}{\frac{1}{5}(0.75)+\frac{4}{5}(0.40)}=0.319$$

Note, the *a priori* probability is 0.20; the outcome raises this probability to 0.319.

10. *Binomial Distribution*

In an experiment consisting of n independent trials in which an event has probability p in a single trial, the probability P_X of obtaining X successes is given by

$$P_X = C_{(n,X)} p^X q^{(n-X)}$$

where

$$q = (1-p) \text{ and } C_{(n,X)} = \frac{n!}{X!(n-X)!}.$$

The probability of between a and b successes (both a and b included) is $P_a + P_{a+1} + \ldots + P_b$, so if $a = 0$ and $b = n$, this sum is

$$\sum_{X=0}^{n} C_{(n,X)} p^X q^{(n-X)} = q^n + C_{(n,1)} q^{n-1} p$$

$$+ C_{(n,2)} q^{n-2} p^2 + \ldots + p^n = (q+p)^n = 1.$$

11. *Mean of Binomially Distributed Variable*

The mean number of successes in n independent trials is $m = np$ with standard deviation $\sigma = \sqrt{npq}$.

12. *Normal Distribution*

In the binomial distribution, as n increases the histogram of heights is approximated by the bell-shaped curve (normal curve)

$$Y = \frac{1}{\sigma\sqrt{2\pi}} e^{-(x-m)^2/2\sigma^2}$$

where m = the mean of the binomial distribution = np, and $\sigma = \sqrt{npq}$ is the standard deviation. For any normally distributed random variable X with mean m and standard deviation σ the probability function (density) is given by the above.

The *standard* normal probability curve is given by

$$y = \frac{1}{\sqrt{2\pi}} e^{-Z^2/2}$$

and has mean = 0 and standard deviation = 1. The total area under the standard normal curve is 1. Any normal variable X can be put into standard form by defining $Z = (X - m)/\sigma$; thus the probability of X between a given X_1 and X_2 is the area under the standard normal curve between the corresponding Z_1 and Z_2 (Table A.1, Appendix). The standard normal curve is often used instead of the binomial distribution in experiments with discrete outcomes. For example, to determine the probability of obtaining 60 to 70 heads in a toss of 100 coins, we take $X = 59.5$ to $X = 70.5$ and compute corresponding values of Z from mean $np = 100 \ \frac{1}{2} = 50$, and the standard deviation $\sigma = \sqrt{(100)(1/2)(1/2)} = 5$. Thus, $Z = (59.5 - 50)/5 = 1.9$ and $Z = (70.5 - 50)/5 = 4.1$. From Table A.1, area between $Z = 0$ and $Z = 4.1$ is 0.5000 and between $Z = 0$ and $Z = 1.9$ is 0.4713; hence, the desired probability is 0.0287. The binomial distribution requires a more lengthy computation

$$C_{(100,60)}(1/2)^{60}(1/2)^{40} + C_{(100,61)}(1/2)^{61}(1/2)^{39} + \ldots + C_{(100,70)}(1/2)^{70}(1/2)^{30}.$$

Note that the normal curve is symmetric, whereas the histogram of the binomial distribution is symmetric only if $p = q = 1/2$. Accordingly, when p (hence q) differ appreciably from 1/2, the difference between probabilities computed by each increases. It is usually recommended that the normal approximation not be used if p (or q) is so small that np (or nq) is less than 5.

13. Poisson Distribution

$$P = \frac{e^{-m} m^r}{r!}$$

is an approximation to the binomial probability for r successes in n trials when $m = np$ is small (<5) and the normal curve is not recommended to approximate binomial probabilities (Table A.2). The variance σ^2 in the Poisson distribution is np, the same value as the mean. ***Example***: A school's expulsion rate is 5 students per 1000. If class size is 400, what is the probability that 3 or more will be expelled? Since $p = 0.005$ and $n = 400$, $m = np = 2$, and $r = 3$. From Table A.2 we obtain for $m = 2$ and $r(=x) = 3$ the probability $p = 0.323$.

14. Empirical Distributions

A distribution that is skewed to the right (positive skewness) has a median to the right of the mode and a mean to the right of the median. One that is negatively skewed has a median to the left of the mode and a mean to the left of the median. An approximate relationship among the three parameters is given by

$$\text{Median} \doteq 2/3\ (\text{mean}) + 1/3\ (\text{mode})$$

Skewness may be measured by either of the formulas:

$$\text{Skewness} = (\text{mean} - \text{mode})/s$$

$$\text{Skewness} = 3(\text{mean} - \text{median})/s$$

15. *Estimation*

Conclusions about a population parameter such as mean μ may be expressed in an interval estimation containing the sample estimate in such a way that the interval includes the unknown μ with probability $(1 - \alpha)$. A value Z_α is obtained from the table for the normal distribution. For example, $Z_\alpha = 1.96$ for $\alpha = 0.05$. Sample values $X_1, X_2, \ldots, X_n$ permit computation of the variance s^2, which is an estimate of σ^2. A confidence interval for μ is

$$(\bar{X} - Z_\alpha s/\sqrt{n}\ ,\ \bar{X} + Z_\alpha s/\sqrt{n}\ )$$

For $\alpha = 0.05$ this interval is

$$(\bar{X} - 1.96 s/\sqrt{n}\ ,\ \bar{X} + 1.96 s/\sqrt{n}\ )$$

The ratio $s/\sqrt{n}$ is the *standard error of the mean* (see Section 17).

16. *Hypotheses Testing*

Two groups may have different sample means and it is desired to know if the apparent difference arises from random or significant deviation in the items of the samples. The *null hypothesis* (H_0) is that both samples belong to the same population, i.e., the differences are random. The alternate hypothesis (H_1) is these are two different populations. Test procedures are designed so one may accept or reject the null hypothesis. The decision to accept is made with probability α of error. The values of α are usually 0.05, 0.01, or 0.001. If the null hypothesis is rejected, though correct, the error is called an *error of the first kind*. The error of acceptance of the null hypothesis, when false, is an *error of the second kind*.

17. *t-Distribution*

In many situations, μ and σ are unknown and must be estimated from $\bar{X}$ and s in a sample of small size n, so use of the normal distribution is not recommended. In such situations the Student t-distribution is used and is given by the probability density function

$$y = A(1 + t^2/f)^{-(f+1)/2}$$

where f stands for degrees of freedom and A is a constant

$$= \Gamma(f/2 + 1/2)/\Gamma(f/2)\sqrt{f\pi}$$

so that the total area (probability) under the curve of y vs. t is 1. In a normally distributed population with mean μ, if all possible samples of size n and mean $\bar{X}$ are taken, the quantity $(\bar{X} - \mu)\sqrt{n}/s$ satisfies the t-distribution with $f = n - 1$, or

$$t=\frac{\overline{X}-\mu}{s/\sqrt{n}}.$$

Thus, confidence limits for μ are

$$(\overline{X}-t\cdot s/\sqrt{n}\,,\overline{X}+t\cdot s/\sqrt{n}\,)$$

where t is obtained from Table A.3 for $(n-1)$ degrees of freedom and confidence level $(1-\alpha)$.

18. Hypothesis Testing with t- and Normal Distributions

When two normal, independent populations with means μ_X and μ_Y and standard deviations σ_X and σ_Y are considered and all possible pairs of samples are taken, the distribution of the difference between sample means $\overline{X}-\overline{Y}$ is also normally distributed. This distribution has mean $\mu_X-\mu_Y$ and standard deviation

$$\sqrt{\frac{\sigma_X^2}{n_1}+\frac{\sigma_Y^2}{n_2}}$$

where n_1 is the sample size of X_i variates and n_2 is the sample size of Y_i variates. The quantity Z computed as

$$Z=\frac{(\overline{X}-\overline{Y})-(\mu_X-\mu_Y)}{\sqrt{\frac{\sigma_X^2}{n_1}+\frac{\sigma_Y^2}{n_2}}}$$

satisfies a standard normal probability curve (Section 11).

Accordingly, to test whether two sample means differ significantly, i.e., whether they are drawn from the same or different populations, the null hypothesis (H_0) is $\mu_X - \mu_Y = 0$, and

$$Z = \frac{\overline{X} - \overline{Y}}{\sqrt{\frac{\sigma_X^2}{n_1} + \frac{\sigma_Y^2}{n_2}}}$$

is computed. For sufficiently large samples ($n_1 > 30$ and $n_2 > 30$), sample standard deviations s_X and s_Y are used as estimates of σ_X and σ_Y, respectively. The difference is significant if the value of Z indicates a small probability, say, < 0.05 (or $|Z| > 1.96$; Table A.1).

For *small samples* where the standard deviation of the population is unknown and estimated from the sample, the t-distribution is used instead of the standard normal curve.

$$t = \frac{(\overline{X} - \overline{Y}) - (\mu_X - \mu_Y)}{\sqrt{\frac{s^2}{n_1} + \frac{s^2}{n_2}}},$$

where s the the "pooled estimate of the standard deviation" computed from

$$s^2 = \frac{(n_1 - 1)s_X^2 + (n_2 - 1)s_Y^2}{n_1 + n_2 - 2}$$

Example: Mean exam scores for 2 groups of students on a standard exam were 75 and 68, with other pertinent values:

$$\bar{X} = 75 \qquad \bar{Y} = 68$$
$$s_x = 4 \qquad s_y = 3$$
$$n_1 = 20 \qquad n_2 = 18$$

Thus,

$$s^2 = \frac{(19)(4)^2 + (17)(3)^2}{36} = 12.7,$$

and

$$t = \frac{75 - 68}{\sqrt{\dfrac{12.7}{20} + \dfrac{12.7}{18}}} = 6.05.$$

From Table A.3, $t_{0.01}$, for 36 degrees of freedom, is between 2.70 and 2.75; hence these means are significantly different at the 0.01 level.

The computed t is compared to the tabular value (Table A.3) for degrees of freedom $f = n_1 + n_2 - 2$ at the appropriate confidence level (such as $\alpha = 0.05$ or 0.01). When the computed t exceeds in magnitude the value from the table, the null hypothesis is rejected and the difference is said to be significant. In cases that involve *pairing* of the variates, such as heart rate before and after exercise, the difference $D = X - Y$ is analyzed. The mean (sample) difference $\bar{D}$ is computed and the null hypothesis is tested from

$$t = \frac{\overline{D}}{s_D/\sqrt{n}},$$

where s_D is the standard deviation of the set of differences:

$$s_D = [\Sigma(D - \overline{D})^2/(n-1)]^{1/2}$$

In this case, $f = n - 1$.

19. *Chi-Square Distribution*

In an experiment with two outcomes (e.g., "heads" or "tails"), the observed frequencies can be compared to the expected frequencies by applying the normal distribution. For more than two outcomes, say n, the observed frequencies $O_1, O_2, \ldots, O_n$ and the expected frequencies, $e_1, e_2, \ldots, e_n$, are compared with the chi-square statistic (χ^2):

$$\chi^2 = \sum_{i=1}^{n} \frac{(O_i - e_i)^2}{e_i}.$$

The χ^2 is well approximated by a theoretical distribution expressed in Table A.4. The probability that χ^2 is between two numbers χ_1^2 and χ_2^2 is the area under the curve between χ_1^2 and χ_2^2 for degrees of freedom f. The probability density function is

$$y = \frac{1}{2^{f/2}\Gamma(f/2)} e^{-\frac{1}{2}\chi^2} (\chi^2)^{(f-2)/2}, \quad (0 \le \chi^2 \le \infty).$$

In a *contingency table* of j rows and k columns, $f = (j-1)(k-1)$. In such a matrix arrangement the observed and expected frequencies are determined for each of the $j \times k = n$ "cells" or positions and entered in the above equation.

Example—Contingency table: Men and women were sampled for preference of three different brands of breakfast cereal. The number of each gender that liked the brand is shown in the contingency table. The expected number for each cell is given in parentheses and is calculated as row total × column total/grand total. Degrees of freedom $= (2-1) \times (3-1) = 2$ and χ^2 is calculated as:

$$\chi^2 = \frac{(50-59.7)^2}{59.7} + \ldots + \frac{(60-75.7)^2}{75.7} = 11.4$$

	Brands			
	A	**B**	**C**	**Totals**
Men	50(59.7)	40(45.9)	80(64.3)	170
Women	80(70.3)	60(54.1)	60(75.7)	200
Totals	130	100	140	370

Since the tabular value at the 5% level for $f = 2$ is 5.99, the result is significant for a relationship between gender and brand preference.

When $f = 1$ the "adjusted" χ^2 formula (Yates' correction) is recommended

$$\chi^2_{\text{adj}} = \sum_{i=1}^{n} \frac{(|O_i - e_i| - 1/2)^2}{e_i}.$$

χ^2 is frequently used to determine whether a population satisfies a normal distribution. A large sample of the population is taken and divided into C classes in each of which the observed frequency is noted and the expected frequency calculated. The latter is calculated from the assumption of a normal distribution. The class intervals should contain an expected frequency of 5 or more. Thus, for the interval (X_i, X_{i+1}) calculations of $Z_i = (X_i - \overline{X})/s$ and $Z_{i+1} = (X_{i+1} - \overline{X})/s$ are made and the probability is determined from the area under the standard normal curve. This probability $(P_i) \times N$ gives the expected frequency for the class interval. Degrees of freedom $= C - 3$ in this application of the χ^2 test.

20. *Least Squares Regression*

A set of n values (X_i, Y_i) that display a linear trend is described by the linear equation $\hat{Y}_i = \alpha + \beta X_i$. Variables α and β are constants (population parameters) and are the intercept and slope, respectively. The rule for determining the line is one minimizing the sum of the squared deviations

$$\sum_{i=1}^{n} (Y_i - \hat{Y}_i)^2$$

and with this *criterion* the parameters α and β are best estimated from a and b calculated as

$$b = \frac{\Sigma X_i Y_i - \dfrac{(\Sigma X_i)(\Sigma Y_i)}{n}}{\Sigma X_i^2 - \dfrac{(\Sigma X_i)^2}{n}}$$

and

$$a = \overline{Y} - b\overline{X},$$

where $\overline{X}$ and $\overline{Y}$ are mean values, assuming that for any value of X the distribution of Y values is normal with variances that are equal for all X and the latter (X) are obtained with negligible error. The null hypothesis, $H_0: \beta = 0$, is tested with analysis of variance:

Source	SS	DF	MS
Total $(Y_i - \overline{Y})$	$\Sigma(Y_i - \overline{Y})^2$	$n-1$	
Regression $(\hat{Y}_i - \overline{Y})$	$\Sigma(\hat{Y}_i - \overline{Y})^2$	1	
Residual $(Y_i - \hat{Y}_i)$	$\Sigma(Y_i - \hat{Y}_i)^2$	$n-2$	$\dfrac{SS_{resid}}{(n-2)} = S^2_{Y \cdot X}$

Computing forms for SS terms are

$$SS_{total} = \Sigma(Y_i - \overline{Y})^2 = \Sigma Y_i^2 - (\Sigma Y_i)^2/n$$

$$SS_{regr.} = \Sigma(\hat{Y}_i - \overline{Y})^2 = \frac{[\Sigma X_i Y_i - (\Sigma X_i)(\Sigma Y_i)/n]^2}{\Sigma X_i^2 - (\Sigma X_i)^2/n}$$

Example: Given points: (0,1), (2,3), (4,9), (5,16). Analysis proceeds with the following calculations. $\Sigma X = 11$; $\Sigma Y = 29$; $\Sigma X^2 = 45$; $\Sigma XY = 122$; $\overline{X} = 2.75$; $\overline{Y} = 7.25$; $b = 2.86$; $\Sigma(X_i - \overline{X})^2 = 14.7 \therefore \hat{Y}_i = -0.615 + 2.86X$.

	SS	DF	MS	
Total	136.7	3		$F = \frac{121}{7.85} = 15.4$ (significant)*
Regr.	121	1	121	
Resid.	15.7	2	$7.85 = S_{Y \cdot X}^2$	$r^2 = 0.885$; $s_b = 0.73$

*(See F-distribution, Section 21.)

$F = MS_{\text{regr.}}/MS_{\text{resid.}}$ is calculated and compared with the critical value of F for the desired confidence level for degrees of freedom 1 and $n-2$ (Table A.5). The coefficient of determination, denoted r^2, is

$$r^2 = SS_{\text{regr.}}/SS_{\text{total}}$$

r is the *correlation coefficient*. The *standard error of estimate* is $\sqrt{s_{Y \cdot X}^2}$ and is used to calculate confidence intervals for α and β. For the confidence limits of β and α

$$b \pm t s_{Y \cdot X} \sqrt{\frac{1}{\Sigma(X_i - \overline{X})^2}}$$

$$a \pm t s_{Y \cdot X} \sqrt{\frac{1}{n} + \frac{\overline{X}^2}{\Sigma(X_i - \overline{X})^2}}$$

where t has $n-2$ degrees of freedom and is obtained from Table A.3 for the required probability.

The null hypothesis $H_0 : \beta = 0$, can also be tested with the t statistic:

$$t = \frac{b}{s_b}$$

where s_b is the standard error of b

$$s_b = \frac{s_{Y \cdot X}}{\left[\Sigma (X_i - \overline{X})^2\right]^{1/2}}$$

- *Standard Error of $\hat{Y}$*

An estimate of the mean value of Y for a given value of X, say X_0, is given by the regression equation

$$\hat{Y}_0 = a + bX_0 .$$

The standard error of this predicted value is given by

$$S_{\hat{Y}_0} = S_{Y \cdot X} \left[\frac{1}{n} + \frac{(X_0 - \overline{X})^2}{\Sigma (X_i - \overline{X})^2} \right]^{\frac{1}{2}}$$

and is a minimum when $X_0 = \overline{X}$ and increases as X_0 moves away from $\overline{X}$ in either direction.

21. The F-Distribution (Analysis of Variance)

Given a normally distributed population from which two independent samples are drawn, these provide estimates, s_1^2 and s_2^2, of the variance σ^2. Quotient

$F = s_1^2/s_2^2$ has this probability density function for f_1 and f_2 degrees of freedom of s_1 and s_2:

$$y = \frac{\Gamma\left(\frac{f_1+f_2}{2}\right)}{\Gamma\left(\frac{f_1}{2}\right)\Gamma\left(\frac{f_2}{2}\right)} \cdot f_1^{\frac{f_1}{2}} f_2^{\frac{f_2}{2}} \cdot \frac{F^{\frac{f_1-2}{2}}}{(f_2+f_1F)^{\frac{f_1+f_2}{2}}}, \; (0 \le f < \infty)$$

In testing among k groups (with sample size n) and sample means $\bar{A}_1, \bar{A}_2, \ldots, \bar{A}_k$, the F-distribution tests the null hypothesis: $\mu_1 = \mu_2 = \ldots = \mu_k$ for the means of populations from which the sample is drawn. Individual values from the jth sample ($j = 1$ to k) are denoted A_{ij} ($i = 1$ to n). The "between means" sums of squares (S.S.T.) is computed

$$\text{S.S.T.} = n(\bar{A}_1 - \bar{A})^2 + n(\bar{A}_2 - \bar{A})^2 + \ldots$$
$$+ n(\bar{A}_k - \bar{A})^2,$$

where $\bar{A}$ is the mean of all group means, as well as the "within-samples" sum of squares (S.S.E.), where

$$\text{S.S.E.} = \sum_{i=1}^{n} (A_{i1} - \bar{A}_1)^2 + \sum_{i=1}^{n} (A_{i2} - \bar{A}_2)^2 + \ldots$$
$$+ \sum_{i=1}^{n} (A_{ik} - \bar{A}_k)^2$$

Then

$$s_1^2 = \frac{\text{S.S.T.}}{k-1}$$

and

$$s_2^2 = \frac{\text{S.S.E.}}{k(n-1)}$$

are calculated and the ratio F is obtained

$$F = \frac{s_1^2}{s_2^2},$$

with numerator degrees of freedom $k-1$ and denominator degrees of freedom $k(n-1)$. If the calculated F exceeds the tabular value of F at the desired probability (say, 0.05) we *reject* the null hypothesis that the samples came from populations with equal means (see Table A.5 and gamma function, Section 9.2).

22. *Summary of Probability Distributions*

• *Continuous Distributions*

Distribution

Normal

$$y = \frac{1}{\sigma\sqrt{2\pi}} \exp[-(x-m)^2/2\sigma^2]$$

Mean $= m$

Variance $= \sigma^2$

Standard normal

$$y = \frac{1}{\sqrt{2\pi}} \exp(-z^2/2)$$

Mean = 0

Variance = 1

F-distribution

$$y = A\frac{F^{\frac{f_1-2}{2}}}{(f_2+f_1F)^{\frac{f_1+f_2}{2}}};$$

$$\text{where } A = \frac{\Gamma\left(\frac{f_1+f_2}{2}\right)}{\Gamma\left(\frac{f_1}{2}\right)\Gamma\left(\frac{f_2}{2}\right)} f_1^{\frac{f_1}{2}} f_2^{\frac{f_2}{2}}$$

$$\text{Mean} = \frac{f_2}{f_2-2}$$

$$\text{Variance} = \frac{2f_2^2(f_1+f_2-2)}{f_1(f_2-2)^2(f_2-4)}$$

Chi-square

$$y = \frac{1}{2^{f/2}\Gamma(f/2)}\exp(-\frac{1}{2}x^2)(x^2)^{\frac{f-2}{2}}$$

Mean $= f$

Variance $= 2f$

Students *t*

$$y = A(1+t^2/f)^{-(f+1)/2}; \text{ where } A = \frac{\Gamma(f/2+1/2)}{\sqrt{f\pi}\,\Gamma(f/2)}$$

Mean $= 0$

Variance $= \dfrac{f}{f-2}$ (for $f > 2$)

• *Discrete Distributions*

Binomial distribution

$y = C_{(n,x)} p^x (1-p)^{n-x}$

Mean $= np$

Variance $= np\,(1-p)$

Poisson distribution

$$y = \frac{e^{-m} m^x}{x!}$$

Mean $= m$

Variance $= m$

Table of Derivatives

In the following table, a and n are constants, e is the base of the natural logarithms, and u and v denote functions of x.

1. $\frac{d}{dx}(a) = 0$

2. $\frac{d}{dx}(x) = 1$

3. $\frac{d}{dx}(au) = a\frac{du}{dx}$

4. $\frac{d}{dx}(u + v) = \frac{du}{dx} + \frac{dv}{dx}$

5. $\frac{d}{dx}(uv) = u\frac{dv}{dx} + v\frac{du}{dx}$

6. $\frac{d}{dx}(u/v) = \frac{v\frac{du}{dx} - u\frac{dv}{dx}}{v^2}$

7. $\frac{d}{dx}(u^n) = nu^{n-1}\frac{du}{dx}$

8. $\frac{d}{dx}e^u = e^u\frac{du}{dx}$

9. $\frac{d}{dx}a^u = (\log_e a)a^u\frac{du}{dx}$

10. $\dfrac{d}{dx}\log_e u = (1/u)\dfrac{du}{dx}$

11. $\dfrac{d}{dx}\log_a u = (\log_a e)(1/u)\dfrac{du}{dx}$

12. $\dfrac{d}{dx}u^v = vu^{v-1}\dfrac{du}{dx} + \mathrm{u}^v(\log_e u)\dfrac{dv}{dx}$

13. $\dfrac{d}{dx}\sin u = \cos u\dfrac{du}{dx}$

14. $\dfrac{d}{dx}\cos u = -\sin u\dfrac{du}{dx}$

15. $\dfrac{d}{dx}\tan u = \sec^2 u\dfrac{du}{dx}$

16. $\dfrac{d}{dx}\operatorname{ctn} u = -\csc^2 u\dfrac{du}{dx}$

17. $\dfrac{d}{dx}\sec u = \sec u \tan u\dfrac{du}{dx}$

18. $\dfrac{d}{dx}\csc u = -\mathrm{csu}\operatorname{ctn} u\dfrac{du}{dx}$

19. $\dfrac{d}{dx}\sin^{-1} u = \dfrac{1}{\sqrt{1-u^2}}\dfrac{du}{dx}, (-\tfrac{1}{2}\pi \leq \sin^{-1} u \leq \tfrac{1}{2}\pi)$

20. $\dfrac{d}{dx}\cos^{-1} u = \dfrac{-1}{\sqrt{1-u^2}}\dfrac{du}{dx}, (0 \leq \cos^{-1} u \leq \pi)$

21. $\dfrac{d}{dx}\tan^{-1} u = \dfrac{1}{1+u^2}\dfrac{du}{dx}$

22. $\dfrac{d}{dx}\operatorname{ctn}^{-1} u = \dfrac{-1}{1+u^2}\dfrac{du}{dx}$

23. $\dfrac{d}{dx}\sec^{-1} u = \dfrac{1}{u\sqrt{u^2-1}}\dfrac{du}{dx}, (-\pi \le \sec^{-1} u < -\tfrac{1}{2}\pi;$

$0 \le \sec^{-1} u < \tfrac{1}{2}\pi)$

24. $\dfrac{d}{dx}\csc^{-1} u = \dfrac{-1}{u\sqrt{u^2-1}}\dfrac{du}{dx}, (-\pi < \csc^{-1} u \le -\tfrac{1}{2}\pi;$

$0 < \csc^{-1} u \le \tfrac{1}{2}\pi)$

25. $\dfrac{d}{dx}\sinh u = \cosh u \dfrac{du}{dx}$

26. $\dfrac{d}{dx}\cosh u = \sinh u \dfrac{du}{dx}$

27. $\dfrac{d}{dx}\tanh u = \operatorname{sech}^2 u \dfrac{du}{dx}$

28. $\dfrac{d}{dx}\operatorname{ctnh} u = -\operatorname{csch}^2 u \dfrac{du}{dx}$

29. $\dfrac{d}{dx}\operatorname{sech} u = -\operatorname{sech} u \tanh u \dfrac{du}{dx}$

30. $\dfrac{d}{dx}\operatorname{csch} u = -\operatorname{csch} u \operatorname{ctnh} u \dfrac{du}{dx}$

31. $\dfrac{d}{dx}\sinh^{-1} u = \dfrac{1}{\sqrt{u^2+1}}\dfrac{du}{dx}$

32. $\dfrac{d}{dx}\cosh^{-1} u = \dfrac{1}{\sqrt{u^2-1}}\dfrac{du}{dx}$

33. $\dfrac{d}{dx}\tanh^{-1} u = \dfrac{1}{1-u^2}\dfrac{du}{dx}$

34. $\dfrac{d}{dx}\operatorname{ctnh}^{-1} u = \dfrac{-1}{u^2-1}\dfrac{du}{dx}$

35. $\dfrac{d}{dx}\operatorname{sech}^{-1} u = \dfrac{-1}{u\sqrt{1-u^2}}\dfrac{du}{dx}$

36. $\dfrac{d}{dx}\operatorname{csch}^{-1} u = \dfrac{-1}{u\sqrt{u^2+1}}\dfrac{du}{dx}$

Additional Relations with Derivatives

$$\frac{d}{dt}\int_a^t f(x)\,dx = f(t)$$

$$\frac{d}{dt}\int_t^a f(x)\,dx = -f(t)$$

If $x = f(y)$, then

$$\frac{dy}{dx} = \frac{1}{\dfrac{dx}{dy}}$$

If $y = f(u)$ and $u = g(x)$, then

$$\frac{dy}{dx} = \frac{dy}{du}\cdot\frac{du}{dx} \qquad \text{(chain rule)}$$

If $x = f(t)$ and $y = g(t)$, then

$$\frac{dy}{dx} = \frac{g'(t)}{f'(t)},$$

and

$$\frac{d^2y}{dx^2} = \frac{f'(t)g''(t) - g'(t)f''(t)}{[f'(t)]^3}$$

(*Note*: exponent in denominator is 3.)

Table of Integrals

Indefinite Integrals
Definite Integrals

Table of Indefinite Integrals

Basic Forms (all logarithms are to base *e*)

1. $\int dx = x + C$

2. $\int x^n \, dx = \frac{x^{n+1}}{n+1} + C, \ (n \neq -1)$

3. $\int \frac{dx}{x} = \log x + C$

4. $\int e^x \, dx = e^x + C$

5. $\int a^x \, dx = \frac{a^x}{\log a} + C$

6. $\int \sin x \, dx = -\cos x + C$

7. $\int \cos x \, dx = \sin x + C$

8. $\int \tan x \, dx = -\log \cos x + C$

9. $\int \sec^2 x \, dx = \tan x + C$

10. $\int \csc^2 x \, dx = -\operatorname{ctn} x + C$

11. $\int \sec x \tan x \, dx = \sec x + c$

12. $\int \sin^2 x \, dx = \frac{1}{2} x - \frac{1}{2} \sin x \cos x + C$

13. $\int \cos^2 x\,dx = \frac{1}{2}x + \frac{1}{2}\sin x \cos x + C$

14. $\int \log x\,dx = x \log x - x + C$

15. $\int a^x \log a\,dx = a^x + C, \quad (a > 0)$

16. $\int \frac{dx}{a^2+x^2} = \frac{1}{a} \arctan \frac{x}{a} + C$

17. $\int \frac{dx}{x^2-a^2} = \frac{1}{2a} \log \frac{x-a}{x+a} + C, (x^2 > a^2)$

$= \frac{1}{2a} \log \frac{a-x}{a+x} + C, (x^2 < a^2)$

18. $\int \frac{dx}{\sqrt{x^2+a^2}} = \log\left(x+\sqrt{x^2+a^2}\right) + C$

19. $\int \frac{dx}{\sqrt{x^2-a^2}} = \log\left(x+\sqrt{x^2-a^2}\right) + C$

20. $\int \frac{dx}{\sqrt{a^2-x^2}} = \arcsin \frac{x}{a} + C$

21. $\int \sqrt{a^2-x^2}\ dx = 1/2\left\{x\sqrt{a^2-x^2} + a^2 \arcsin \frac{x}{a}\right\} + C$

22. $\int \sqrt{a^2+x^2}\ dx = 1/2\left\{x\sqrt{a^2+x^2} + a^2 \log(x+\sqrt{a^2+x^2})\right\} + C$

23. $\int \sqrt{x^2-a^2}\ dx = 1/2\left\{x\sqrt{x^2-a^2} - a^2 \log(x+\sqrt{x^2-a^2})\right\} + C$

In the following list, a constant of integration C should be added to the result of each integration.

Form $ax + b$

24. $$\int (ax+b)^m\,dx = \frac{(ax+b)^{m+1}}{a(m+1)}$$

$(m \neq -1)$

25. $$\int x(ax+b)^m\,dx = \frac{(ax+b)^{m+2}}{a^2(m+2)} - \frac{b(ax+b)^{m+1}}{a^2(m+1)},$$

$(m \neq -1, -2)$

26. $$\int \frac{dx}{ax+b} = \frac{1}{a}\log(ax+b)$$

27. $$\int \frac{dx}{(ax+b)^2} = -\frac{1}{a(ax+b)}$$

28. $$\int \frac{dx}{(ax+b)^3} = -\frac{1}{2a(ax+b)^2}$$

29. $$\int \frac{x\,dx}{ax+b} = \frac{x}{a} - \frac{b}{a^2}\log(ax+b)$$

30. $$\int \frac{x\,dx}{(ax+b)^2} = \frac{b}{a^2(ax+b)} + \frac{1}{a^2}\log(ax+b)$$

31. $$\int \frac{x\,dx}{(ax+b)^3} = \frac{b}{2a^2(ax+b)^2} - \frac{1}{a^2(ax+b)}$$

32. $$\int x^2(ax+b)^m\,dx$$

$$= \frac{1}{a^3}\left[\frac{(ax+b)^{m+3}}{m+3} - \frac{2b(ax+b)^{m+2}}{m+2} + \frac{b^2(ax+b)^{m+1}}{m+1}\right]$$

$(m \neq -1, -2, -3)$

33. $$\int \frac{x^2\,dx}{ax+b} = \frac{1}{a^3}\left[\tfrac{1}{2}(ax+b)^2 - 2b(ax+b) + b^2 \log(ax+b)\right]$$

34. $$\int \frac{x^2\,dx}{(ax+b)^2} = \frac{1}{a^3}\left[(ax+b) - \frac{b^2}{ax+b} - 2b\log(ax+b)\right]$$

35. $$\int \frac{x^2\,dx}{(ax+b)^3} = \frac{1}{a^3}\left[\log(ax+b) + \frac{2b}{ax+b} - \frac{b^2}{2(ax+b)^2}\right]$$

36. $$\int \frac{dx}{x(ax+b)} = \frac{1}{b}\log\left(\frac{x}{ax+b}\right)$$

37. $$\int \frac{dx}{x^2(ax+b)} = -\frac{1}{bx} + \frac{a}{b^2}\log\left(\frac{ax+b}{x}\right)$$

38. $$\int \frac{dx}{x(ax+b)^2} = \frac{1}{b(ax+b)} - \frac{1}{b^2}\log\left(\frac{ax+b}{x}\right)$$

39. $$\int \frac{dx}{x^2(ax+b)^2} = -\frac{2ax+b}{b^2x(ax+b)} + \frac{2a}{b^3}\log\left(\frac{ax+b}{x}\right)$$

$$= \frac{1}{a(m+n+1)}\left[x^m(ax+b)^{n+1} - mb\int x^{m-1}(ax+b)^n\,dx\right]$$

$$= \frac{1}{m+n+1}\left[x^{m+1}(ax+b)^n + nb\int x^m(ax+b)^{n-1}\,dx\right]$$

$(m > 0, \quad m+n+1 \neq 0)$

Forms $ax + b$ and $cx + d$

41. $$\int \frac{dx}{(ax+b)(cx+d)} = \frac{1}{bc-ad}\log\left(\frac{cx+d}{ax+b}\right)$$

42. $$\int \frac{x\,dx}{(ax+b)(cx+d)} = \frac{1}{bc-ad}\left[\frac{b}{a}\log(ax+b) - \frac{d}{c}\log(cx+d)\right]$$

43. $$\int \frac{dx}{(ax+b)^2(cx+d)} = \frac{1}{bc-ad}\left[\frac{1}{ax+b} + \frac{c}{bc-ad}\log\left(\frac{cx+d}{ax+b}\right)\right]$$

44. $$\int \frac{x\,dx}{(ax+b)^2(cx+d)}$$

$$= \frac{1}{bc-ad}\left[-\frac{b}{a(ax+b)} - \frac{d}{bc-ad}\log\left(\frac{cx+d}{ax+b}\right)\right]$$

45. $$\int \frac{x^2\,dx}{(ax+b)^2(cx+d)} = \frac{b^2}{a^2(bc-ad)(ax+b)} + \frac{1}{(bc-ad)^2}\left[\frac{d^2}{c}\log|cx+d| + \frac{b(bc-2ad)}{a^2}\log\left|ax+b\right|\right]$$

46. $$\int \frac{ax+b}{cx+d}\,dx = \frac{ax}{c} + \frac{bc-ad}{c^2}\log(cx+d)$$

47. $$\int (ax+b)^m(cx+d)^n\,dx = \frac{1}{a(m+n+1)}\left[(ax+b)^{m+1}(cx+d)^n - n(bc-ad)\int (ax+b)^m(cx+d)^{n-1}\,dx\right]$$

Forms with $\sqrt{ax+b}$

48. $$\int \sqrt{ax+b}\,dx = \frac{2}{3a}\sqrt{(ax+b)^3}$$

49. $$\int x\sqrt{ax+b}\,dx = \frac{2(3ax-2b)}{15a^2}\sqrt{(ax+b)^3}$$

50. $$\int x^2\sqrt{ax+b}\,dx = \frac{2(15a^2x^2 - 12abx + 8b^2)}{105a^3}\sqrt{(ax+b)^3}$$

51. $$\int x^m\sqrt{ax+b}\,dx = \frac{2}{a(2m+3)}\left[x^m\sqrt{(ax+b)^3} - mb\int x^{m-1}\sqrt{ax+b}\,dx\right]$$

52. $$\int \frac{(ax+b)^{\frac{m}{2}}\,dx}{x} = a\int (ax+b)^{\frac{m-2}{2}}\,dx + b\int \frac{(ax+b)^{\frac{m-2}{2}}\,dx}{x}$$

53. $$\int \frac{dx}{x(ax+b)^{\frac{m}{2}}} = \frac{1}{b}\int \frac{dx}{x(ax+b)^{\frac{m-2}{2}}} - \frac{a}{b}\int \frac{dx}{(ax+b)^{\frac{m}{2}}}$$

54. $$\int \frac{\sqrt{ax+b}\,dx}{cx+d} = \frac{2\sqrt{ax+b}}{c} + \frac{1}{c}\sqrt{\frac{bc-ad}{c}}\log\left|\frac{\sqrt{c(ax+b)} - \sqrt{bc-ad}}{\sqrt{c(ax+b)} + \sqrt{bc-ad}}\right|$$

$(c > 0, \quad bc > ad)$

55. $$\int \frac{\sqrt{ax+b}\,dx}{cx+d} = \frac{2\sqrt{ax+b}}{c} - \frac{2}{c}\sqrt{\frac{ad-bc}{c}}\arctan\sqrt{\frac{c(ax+b)}{ad-bc}}$$

$(c > 0, \quad bc < ad)$

56. $$\int \frac{(cx+d)\,dx}{\sqrt{ax+b}} = \frac{2}{3a^2}(3ad - 2bc + acx)\sqrt{ax+b}$$

57. $$\int \frac{dx}{(cx+d)\sqrt{ax+b}} = \frac{2}{\sqrt{c}\sqrt{ad-bc}} \arctan \sqrt{\frac{c(ax+b)}{ad-bc}}$$

$$(c > 0, \quad bc < ad)$$

58. $$\int \frac{dx}{(cx+d)\sqrt{ax+b}} = \frac{1}{\sqrt{c}\sqrt{bc-ad}} \log \left| \frac{\sqrt{c(ax+b)} - \sqrt{bc-ad}}{\sqrt{c(ax+b)} + \sqrt{bc-ad}} \right|$$

$$(c > 0, \quad bc > ad)$$

59. $$\int \sqrt{ax+b}\,\sqrt{cx+d}\,dx = \int \sqrt{acx^2 + (ad+bc)x + bd}\,dx$$

(see 154)

60. $$\int \frac{\sqrt{ax+b}\,dx}{x} = 2\sqrt{ax+b} + \sqrt{b}\log\left(\frac{\sqrt{ax+b} - \sqrt{b}}{\sqrt{ax+b} + \sqrt{b}}\right) \quad (b > 0)$$

61. $$\int \frac{\sqrt{ax+b}\,dx}{x} = 2\sqrt{ax+b} - 2\sqrt{-b}\arctan\sqrt{\frac{ax+b}{-b}} \quad (b < 0)$$

62. $$\int \frac{\sqrt{ax+b}}{x^2}\,dx = -\frac{\sqrt{ax+b}}{x} + \frac{a}{2}\int \frac{dx}{x\sqrt{ax+b}}$$

63. $\int \frac{\sqrt{ax+b}\,dx}{x^m} = -\frac{1}{(m-1)b}\left[\frac{\sqrt{(ax+b)^3}}{x^{m-1}} + \frac{(2m-5)a}{2}\int \frac{\sqrt{ax+b}\,dx}{x^{m-1}}\right]$
$(m \neq 1)$

64. $\int \frac{dx}{\sqrt{ax+b}} = \frac{2\sqrt{ax+b}}{a}$

65. $\int \frac{x\,dx}{\sqrt{ax+b}} = \frac{2(ax-2b)\sqrt{ax+b}}{3a^2}$

66. $\int \frac{x^2\,dx}{\sqrt{ax+b}} = \frac{2(3a^2x^2 - 4abx + 8b^2)\sqrt{ax+b}}{15a^3}$

67. $\int \frac{x^m\,dx}{\sqrt{ax+b}} = \frac{2}{a(2m+1)}\left[x^m\sqrt{ax+b} - mb\int \frac{x^{m-1}\,dx}{\sqrt{ax+b}}\right] \qquad (m \neq -\tfrac{1}{2})$

68. $\int \frac{dx}{x\sqrt{ax+b}} = \frac{1}{\sqrt{b}}\log\left|\frac{\sqrt{ax+b}-\sqrt{b}}{\sqrt{ax+b}+\sqrt{b}}\right| \qquad (b > 0)$

69. $\int \frac{dx}{x\sqrt{ax+b}} = \frac{2}{\sqrt{-b}}\arctan\sqrt{\frac{ax+b}{-b}} \qquad (b < 0)$

70. $$\int \frac{dx}{x^2\sqrt{ax+b}} = -\frac{\sqrt{ax+b}}{bx} - \frac{a}{2b}\int \frac{dx}{x\sqrt{ax+b}}$$

71. $$\int \frac{dx}{x^m\sqrt{ax+b}} = -\frac{\sqrt{ax+b}}{(m-1)bx^{m-1}} - \frac{(2m-3)a}{(2m-2)b}\int \frac{dx}{x^{m-1}\sqrt{ax+b}} \qquad (m \neq 1)$$

72. $$\int (ax+b)^{\pm\frac{m}{2}}\, dx = \frac{2(ax+b)^{\frac{2\pm m}{2}}}{a(2 \pm m)}$$

73. $$\int x(ax+b)^{\pm\frac{m}{2}}\, dx = \frac{2}{a^2}\left[\frac{(ax+b)^{\frac{4\pm m}{2}}}{4 \pm m} - \frac{b(ax+b)^{\frac{2\pm m}{2}}}{2 \pm m}\right]$$

Form $ax^2 + c$

74. $$\int \frac{dx}{ax^2+c} = \frac{1}{\sqrt{ac}} \arctan\left(x\sqrt{\frac{a}{c}}\right) \qquad (a > 0, \quad c > 0)$$

75. $$\int \frac{dx}{ax^2+c} = \frac{1}{2\sqrt{-ac}} \log\left(\frac{x\sqrt{a} - \sqrt{-c}}{x\sqrt{a} + \sqrt{-c}}\right) \qquad (a > 0, \quad c < 0)$$

76. $$\int \frac{dx}{ax^2 + c} = \frac{1}{2\sqrt{-ac}} \log \left(\frac{\sqrt{c} + x\sqrt{-a}}{\sqrt{c} - x\sqrt{-a}} \right) \qquad (a < 0, \quad c > 0)$$

77. $$\int \frac{x\,dx}{ax^2 + c} = \frac{1}{2a} \log (ax^2 + c)$$

78. $$\int \frac{x^2\,dx}{ax^2 + c} = \frac{x}{a} - \frac{c}{a} \int \frac{dx}{ax^2 + c}$$

79. $$\int \frac{x^m\,dx}{ax^2 + c} = \frac{x^{m-1}}{a(m-1)} - \frac{c}{a} \int \frac{x^{m-2}\,dx}{ax^2 + c} \qquad (m \neq 1)$$

80. $$\int \frac{dx}{x(ax^2 + c)} = \frac{1}{2c} \log \left(\frac{ax^2}{ax^2 + c} \right)$$

81. $$\int \frac{dx}{x^2(ax^2 + c)} = -\frac{1}{cx} - \frac{a}{c} \int \frac{dx}{ax^2 + c}$$

82. $$\int \frac{dx}{x^m(ax^2 + c)} = -\frac{1}{c(m-1)x^{m-1}} - \frac{a}{c} \int \frac{dx}{x^{m-2}(ax^2 + c)} \qquad (m \neq 1)$$

83. $$\int \frac{dx}{(ax^2+c)^m} = \frac{1}{2(m-1)c} \cdot \frac{x}{(ax^2+c)^{m-1}} + \frac{2m-3}{2(m-1)c} \int \frac{dx}{(ax^2+c)^{m-1}} \qquad (m \neq 1)$$

84. $$\int \frac{x\,dx}{(ax^2+c)^m} = -\frac{1}{2a(m-1)(ax^2+c)^{m-1}} \qquad (m \neq 1)$$

85. $$\int \frac{x^2\,dx}{(ax^2+c)^m} = -\frac{x}{2a(m-1)(ax^2+c)^{m-1}} + \frac{1}{2a(m-1)}\int \frac{dx}{(ax^2+c)^{m-1}} \qquad (m \neq 1)$$

86. $$\int \frac{dx}{x(ax^2+c)^m} = \frac{1}{2c(m-1)(ax^2+c)^{m-1}} + \frac{1}{c}\int \frac{dx}{x(ax^2+c)^{m-1}} \qquad (m \neq 1)$$

87. $$\int \frac{dx}{x^2(ax^2+c)^m} = \frac{1}{c}\int \frac{dx}{x^2(ax^2+c)^{m-1}} - \frac{a}{c}\int \frac{dx}{(ax^2+c)^m}$$

(see 82 and 83)

Form $ax^2 + bx + c$

88. $$\int \frac{dx}{ax^2+bx+c} = \frac{1}{\sqrt{b^2-4ac}} \log \frac{2ax+b-\sqrt{b^2-4ac}}{2ax+b+\sqrt{b^2-4ac}} \qquad (b^2 > 4ac)$$

89. $$\int \frac{dx}{ax^2+bx+c} = \frac{2}{\sqrt{4ac-b^2}} \tan^{-1}\frac{2ax+b}{\sqrt{4ac-b^2}} \qquad (b^2 < 4ac)$$

90. $$\int \frac{dx}{ax^2 + bx + c} = -\frac{2}{2ax + b}, \ (b^2 = 4ac.)$$

91. $$\int \frac{dx}{(ax^2 + bx + c)^{n+1}} = \frac{2ax + b}{n(4ac - b^2)(ax^2 + bx + c)^n} + \frac{2(2n - 1)a}{n(4ac - b^2)} \int \frac{dx}{(ax^2 + bx + c)^n}$$

92. $$\int \frac{xdx}{ax^2+bx+c} = \frac{1}{2a} \log (ax^2+bx+c) - \frac{b}{2a} \int \frac{dx}{ax^2 - bx + c}$$

93. $$\int \frac{x^2dx}{ax^2+bx+c} = \frac{x}{a} - \frac{b}{2a^2} \log(ax^2 + bx + c) + \frac{b^2 - 2ac}{2a^2} \int \frac{dx}{ax^2 + bx + c}$$

94. $$\int \frac{x^n dx}{ax^2 + bx + c} = \frac{x^{n-1}}{(n - 1)a} - \frac{c}{a} \int \frac{x^{n-2}dx}{ax^2 + bx + c} - \frac{b}{a} \int \frac{x^{n-1}dx}{ax^2 + bx + c}$$

95. $$\int \frac{x\,dx}{(ax^2 + bx + c)^{n+1}} = \frac{-(2c + bx)}{n(4ac - b^2)(ax^2 + bx + c)^n} - \frac{b(2n - 1)}{n(4ac - b^2)} \int \frac{dx}{(ax^2 + bx + c)^n}$$

96. $$\int \frac{dx}{x(ax^2+bx+c)} = \frac{1}{2c} \log\left(\frac{x^2}{ax^2+bx+c}\right) - \frac{b}{2c} \int \frac{dx}{(ax^2+bx+c)}$$

97. $$\int \frac{dx}{x^2(ax^2 + bx + c)} = \frac{b}{2c^2} \log\left(\frac{ax^2 + bx + c}{x^2}\right) - \frac{1}{cx} + \left(\frac{b^2}{2c^2} - \frac{a}{c}\right) \int \frac{dx}{(ax^2 + bx + c)}$$

Forms with $\sqrt{2ax - x^2}$

98. $$\int \sqrt{2ax - x^2}\,dx = \frac{x - a}{2}\sqrt{2ax - x^2} + \frac{a^2}{2} \arcsin\left(\frac{x - a}{a}\right)$$

99. $$\int x\sqrt{2ax - x^2}\,dx = -\frac{3a^2 + ax - 2x^2}{6}\sqrt{2ax - x^2} + \frac{a^3}{2} \arcsin\left(\frac{x - a}{a}\right)$$

100. $$\int x^m\sqrt{2ax - x^2}\,dx$$

$$= -\frac{x^{m-1}\sqrt{(2ax - x^2)^3}}{m + 2} + \frac{a(2m + 1)}{m + 2}\int x^{m-1}\sqrt{2ax - x^2}\,dx$$

101. $$\int \frac{\sqrt{2ax - x^2}\,dx}{x} = \sqrt{2ax - x^2} + a\arcsin\left(\frac{x - a}{a}\right)$$

102. $$\int \frac{\sqrt{2ax - x^2}\,dx}{x^m} = -\frac{\sqrt{(2ax - x^2)^3}}{a(2m - 3)x^m} + \frac{m - 3}{a(2m - 3)}\int \frac{\sqrt{2ax - x^2}\,dx}{x^{m-1}}$$

103. $$\int \frac{dx}{\sqrt{2ax - x^2}} = \arcsin\left(\frac{x - a}{a}\right)$$

104. $$\int \frac{x\,dx}{\sqrt{2ax - x^2}} = -\sqrt{2ax - x^2} + a\arcsin\left(\frac{x - a}{a}\right)$$

105. $$\int \frac{x^m\,dx}{\sqrt{2ax - x^2}} = -\frac{x^{m-1}\sqrt{2ax - x^2}}{m} + \frac{a(2m - 1)}{m}\int \frac{x^{m-1}\,dx}{\sqrt{2ax - x^2}}$$

106. $\int \frac{dx}{x\sqrt{2ax - x^2}} = -\frac{\sqrt{2ax - x^2}}{ax}$

107. $\int \frac{dx}{x^m\sqrt{2ax - x^2}} = -\frac{\sqrt{2ax - x^2}}{a(2m-1)x^m} + \frac{m-1}{a(2m-1)}\int \frac{dx}{x^{m-1}\sqrt{2ax - x^2}}$

Forms with $\sqrt{a^2 - x^2}$

108. $\int \sqrt{a^2 - x^2}\, dx = \tfrac{1}{2}\left(x\sqrt{a^2 - x^2} + a^2 \arcsin \frac{x}{a}\right)$

109. $\int x\sqrt{a^2 - x^2}\, dx = -\tfrac{1}{3}\sqrt{(a^2 - x^2)^3}$

110. $\int x^2\sqrt{a^2 - x^2}\, dx = -\frac{x}{4}\sqrt{(a^2 - x^2)^3} + \frac{a^2}{8}\left(x\sqrt{a^2 - x^2} + a^2 \arcsin \frac{x}{a}\right)$

111. $\int x^3\sqrt{a^2 - x^2}\, dx = \left(-\tfrac{1}{5}x^2 - \tfrac{2}{15}a^2\right)\sqrt{(a^2 - x^2)^3}$

112. $\displaystyle\int \frac{\sqrt{a^2 - x^2}\,dx}{x} = \sqrt{a^2 - x^2} - a \log \left| \frac{a + \sqrt{a^2 - x^2}}{x} \right|$

113. $\displaystyle\int \frac{\sqrt{a^2 - x^2}\,dx}{x^2} = -\frac{\sqrt{a^2 - x^2}}{x} - \arcsin \frac{x}{a}$

114. $\displaystyle\int \frac{\sqrt{a^2 - x^2}\,dx}{x^3} = -\frac{\sqrt{a^2 - x^2}}{2x^2} + \frac{1}{2a} \log \left| \frac{a + \sqrt{a^2 - x^2}}{x} \right|$

115. $\displaystyle\int \frac{dx}{\sqrt{a^2 - x^2}} = \arcsin \frac{x}{a}$

116. $\displaystyle\int \frac{x\,dx}{\sqrt{a^2 - x^2}} = -\sqrt{a^2 - x^2}$

117. $\displaystyle\int \frac{x^2\,dx}{\sqrt{a^2 - x^2}} = -\frac{x}{2}\sqrt{a^2 - x^2} + \frac{a^2}{2} \arcsin \frac{x}{a}$

118. $\displaystyle\int \frac{x^3\,dx}{\sqrt{a^2 - x^2}} = \tfrac{1}{3}\sqrt{(a^2 - x^2)^3} - a^2\sqrt{a^2 - x^2}$

119. $$\int \frac{dx}{x\sqrt{a^2 - x^2}} = -\frac{1}{a}\log\left|\frac{a + \sqrt{a^2 - x^2}}{x}\right|$$

120. $$\int \frac{dx}{x^2\sqrt{a^2 - x^2}} = -\frac{\sqrt{a^2 - x^2}}{a^2x}$$

121. $$\int \frac{dx}{x^3\sqrt{a^2 - x^2}} = -\frac{\sqrt{a^2 - x^2}}{2a^2x^2} - \frac{1}{2a^3}\log\left|\frac{a + \sqrt{a^2 - x^2}}{x}\right|$$

Forms with $\sqrt{x^2 - a^2}$

122. $$\int \sqrt{x^2 - a^2}\,dx = \frac{x}{2}\sqrt{x^2 - a^2} - \frac{a^2}{2}\log\left|x + \sqrt{x^2 - a^2}\right|$$

123. $$\int x\sqrt{x^2 - a^2}\,dx = \tfrac{1}{3}\sqrt{(x^2 - a^2)^3}$$

124. $$\int x^2\sqrt{x^2 - a^2}\,dx = \frac{x}{4}\sqrt{(x^2 - a^2)^3} + \frac{a^2x}{8}\sqrt{x^2 - a^2} - \frac{a^4}{8}\log\left|x + \sqrt{x^2 - a^2}\right|$$

125. $$\int x^3\sqrt{x^2 - a^2}\,dx = \frac{1}{5}\sqrt{(x^2 - a^2)^5} + \frac{a^2}{3}\sqrt{(x^2 - a^2)^3}$$

126. $\int \frac{\sqrt{x^2 - a^2}\,dx}{x} = \sqrt{x^2 - a^2} - a \arccos \frac{a}{x}$

127. $\int \frac{\sqrt{x^2 - a^2}\,dx}{x^2} = -\frac{1}{x}\sqrt{x^2 - a^2} + \log |x + \sqrt{x^2 -$

128. $\int \frac{\sqrt{x^2 - a^2}\,dx}{x^3} = -\frac{\sqrt{x^2 - a^2}}{2x^2} + \frac{1}{2a} \arccos \frac{a}{x}$

129. $\int \frac{dx}{\sqrt{x^2 - a^2}} = \log |x + \sqrt{x^2 - a^2}|$

130. $\int \frac{x\,dx}{\sqrt{x^2 - a^2}} = \sqrt{x^2 - a^2}$

131. $\int \frac{x^2\,dx}{\sqrt{x^2 - a^2}} = \frac{x}{2}\sqrt{x^2 - a^2} + \frac{a^2}{2} \log |x + \sqrt{x^2 - a^2}|$

132. $\int \frac{x^3\,dx}{\sqrt{x^2 - a^2}} = \frac{1}{3}\sqrt{(x^2 - a^2)^3} + a^2\sqrt{x^2 - a^2}$

133. $$\int \frac{dx}{x\sqrt{x^2 - a^2}} = \frac{1}{a} \arccos \frac{a}{x}$$

134. $$\int \frac{dx}{x^2\sqrt{x^2 - a^2}} = \frac{\sqrt{x^2 - a^2}}{a^2 x}$$

135. $$\int \frac{dx}{x^3\sqrt{x^2 - a^2}} = \frac{\sqrt{x^2 - a^2}}{2a^2x^2} + \frac{1}{2a^3} \arccos \frac{a}{x}$$

Forms with $\sqrt{a^2 + x^2}$

136. $$\int \sqrt{a^2 + x^2}\, dx = \frac{x}{2}\sqrt{a^2 + x^2} + \frac{a^2}{2} \log\left(x + \sqrt{a^2 + x^2}\right)$$

137. $$\int x\sqrt{a^2 + x^2}\, dx = \tfrac{1}{3}\sqrt{(a^2 + x^2)^3}$$

138. $$\int x^2\sqrt{a^2+x^2}\, dx = \frac{x}{4}\sqrt{(a^2+x^2)^3} - \frac{a^2 x}{8}\sqrt{a^2+x^2} - \frac{a^4}{8} \log\left(x + \sqrt{a^2+x^2}\right)$$

139. $\int x^3\sqrt{a^2+x^2}\,dx = (\tfrac{1}{5}x^2 - \tfrac{2}{15}a^2)\sqrt{(a^2+x^2)^3}$

140. $\int \frac{\sqrt{a^2+x^2}\,dx}{x} = \sqrt{a^2+x^2} - a\log\left|\frac{a+\sqrt{a^2+x^2}}{x}\right|$

141. $\int \frac{\sqrt{a^2+x^2}\,dx}{x^2} = -\frac{\sqrt{a^2+x^2}}{x} + \log(x+\sqrt{a^2+x^2})$

142. $\int \frac{\sqrt{a^2+x^2}\,dx}{x^3} = -\frac{\sqrt{a^2+x^2}}{2x^2} - \frac{1}{2a}\log\left|\frac{a+\sqrt{a^2+x^2}}{x}\right|$

143. $\int \frac{dx}{\sqrt{a^2+x^2}} = \log(x+\sqrt{a^2+x^2})$

144. $\int \frac{x\,dx}{\sqrt{a^2+x^2}} = \sqrt{a^2+x^2}$

145. $\int \frac{x^2\,dx}{\sqrt{a^2+x^2}} = \frac{x}{2}\sqrt{a^2+x^2} - \frac{a^2}{2}\log(x+\sqrt{a^2+x^2})$

146. $$\int \frac{x^3\,dx}{\sqrt{a^2+x^2}} = \tfrac{1}{3}\sqrt{(a^2+x^2)^3} - a^2\sqrt{a^2+x^2}$$

147. $$\int \frac{dx}{x\sqrt{a^2+x^2}} = -\frac{1}{a}\log\left|\frac{a+\sqrt{a^2+x^2}}{x}\right|$$

148. $$\int \frac{dx}{x^2\sqrt{a^2+x^2}} = -\frac{\sqrt{a^2+x^2}}{a^2x}$$

149. $$\int \frac{dx}{x^3\sqrt{a^2+x^2}} = -\frac{\sqrt{a^2+x^2}}{2a^2x^2} + \frac{1}{2a^3}\log\left|\frac{a+\sqrt{a^2+x^2}}{x}\right|$$

Forms with $\sqrt{ax^2+bx^2+c}$

150. $$\int \frac{dx}{\sqrt{ax^2+bx+c}} = \frac{1}{\sqrt{a}}\log(2ax+b+2\sqrt{a}\sqrt{ax^2+bx+c}),\ \ a>0.$$

151. $$\int \frac{dx}{\sqrt{ax^2+bx+c}} = \frac{1}{\sqrt{-a}}\sin^{-1}\frac{-2ax-b}{\sqrt{b^2-4ac}},\ \ a<0.$$

152. $$\int \frac{x\,dx}{\sqrt{ax^2+bx+c}} = \frac{\sqrt{ax^2+bx+c}}{a} - \frac{b}{2a}\int\frac{dx}{\sqrt{ax^2+bx+c}}$$

153. $$\int \frac{x^n dx}{\sqrt{ax^2 + bx + c}} = \frac{x^{n-1}}{an}\sqrt{ax^2 + bx + c} - \frac{b(2n-1)}{2an}\int \frac{x^{n-1}dx}{\sqrt{ax^2 + bx + c}} - \frac{c(n-1)}{an}\int \frac{x^{n-2}dx}{\sqrt{ax^2 + bx + c}}$$

154. $$\int \sqrt{ax^2 + bx + c}\, dx = \frac{2ax + b}{4a}\sqrt{ax^2 + bx + c} + \frac{4ac - b^2}{8a}\int \frac{dx}{\sqrt{ax^2 + bx + c}}$$

155. $$\int x\sqrt{ax^2+bx+c}\, dx = \frac{(ax^2+bx+c)^{\frac{3}{2}}}{3a} - \frac{b}{2a}\int \sqrt{ax^2+bx+c}\, dx.$$

156. $$\int x^2\sqrt{ax^2 + bx + c}\, dx = \left(x - \frac{5b}{6a}\right)\frac{(ax^2 + bx + c)^{\frac{3}{2}}}{4a} + \frac{(5b^2 - 4ac)}{16a^2}\int \sqrt{ax^2 + bx + c}\, dx.$$

157. $$\int \frac{dx}{x\sqrt{ax^2+bx+c}} = -\frac{1}{\sqrt{c}} \log\left(\frac{\sqrt{ax^2+bx+c}+\sqrt{c}}{x}+\frac{b}{2\sqrt{c}}\right), \ c>0.$$

158. $$\int \frac{dx}{x\sqrt{ax^2+bx+c}} = \frac{1}{\sqrt{-c}} \sin^{-1} \frac{bx+2c}{x\sqrt{b^2-4ac}}, \quad c<0.$$

159. $$\int \frac{dx}{x\sqrt{ax^2+bx}} = -\frac{2}{bx}\sqrt{ax^2+bx}, \quad c=0.$$

160. $$\int \frac{dx}{x^n\sqrt{ax^2+bx+c}} = -\frac{\sqrt{ax^2+bx+c}}{c(n-1)x^{n-1}}$$
$$+\frac{b(3-2n)}{2c(n-1)}\int \frac{dx}{x^{n-1}\sqrt{ax^2+bx+c}} + \frac{a(2-n)}{c(n-1)}\int \frac{dx}{x^{n-2}\sqrt{ax^2+bx+c}}.$$

161. $$\int \frac{dx}{(ax^2+bx+c)^{\frac{3}{2}}} = -\frac{2(2ax+b)}{(b^2-4ac)\sqrt{ax^2+bx+c}}, \quad b^2 \neq 4ac.$$

162. $$\int \frac{dx}{(ax^2+bx+c)^{\frac{3}{2}}} = -\frac{1}{2\sqrt{a^3}(x+b/2a)^2}, \quad b^2 = 4ac.$$

Miscellaneous Algebraic Forms

163. $$\int \sqrt{\frac{a+x}{b+x}}\,dx = \sqrt{(a+x)(b+x)} + (a-b)\log\left(\sqrt{a+x}+\sqrt{b+x}\right)$$

$$(a + x > 0 \text{ and } b + x > 0)$$

164. $$\int \sqrt{\frac{a+x}{b-x}}\,dx = -\sqrt{(a+x)(b-x)} - (a+b)\arcsin\sqrt{\frac{b-x}{a+b}}$$

165. $$\int \sqrt{\frac{a-x}{b+x}}\,dx = \sqrt{(a-x)(b+x)} + (a+b)\arcsin\sqrt{\frac{b+x}{a+b}}$$

166. $$\int \sqrt{\frac{1+x}{1-x}}\,dx = -\sqrt{1-x^2} + \arcsin x$$

167. $$\int \frac{dx}{\sqrt{(x-a)(b-x)}} = 2\arcsin\sqrt{\frac{x-a}{b-a}}$$

168. $$\int \frac{dx}{ax^3+b} = \frac{k}{3b}\left[\sqrt{3}\arctan\frac{2x-k}{k\sqrt{3}} + \log\left|\frac{k+x}{\sqrt{x^2-kx+k^2}}\right|\right]$$

$$\left(b \neq 0,\ k = \sqrt[3]{\frac{b}{a}}\right)$$

169. $$\int \frac{x\,dx}{ax^3 + b} = \frac{1}{3ak}\left[\sqrt{3}\arctan\frac{2x - k}{k\sqrt{3}} - \log\left|\frac{k + x}{\sqrt{x^2 - kx + k^2}}\right|\right]$$
$$\left(b \neq 0,\ k = \sqrt[3]{\frac{a}{b}}\right)$$

170. $$\int \frac{dx}{x(ax^m + b)} = \frac{1}{bm}\log\left|\frac{x^m}{ax^m + b}\right| \qquad (b \neq 0)$$

171. $$\int \frac{dx}{\sqrt{(2ax - x^2)^3}} = \frac{x - a}{a^2\sqrt{2ax - x^2}}$$

172. $$\int \frac{x\,dx}{\sqrt{(2ax - x^2)^3}} = \frac{x}{a\sqrt{2ax - x^2}}$$

173. $$\int \frac{dx}{\sqrt{2ax + x^2}} = \log\left|x + a + \sqrt{2ax + x^2}\right|$$

174. $$\int \sqrt{\frac{cx + d}{ax + b}}\,dx = \frac{\sqrt{ax + b}\cdot\sqrt{cx + d}}{a} + \frac{(ad - bc)}{2a}\int \frac{dx}{\sqrt{ax + b}\cdot\sqrt{cx + d}}$$

Trigonometric Forms

175. $\int (\sin ax)\, dx = -\frac{1}{a}\cos ax$

176 $\int (\sin^2 ax)\, dx = -\frac{1}{2a}\cos ax \sin ax + \frac{1}{2}x = \frac{1}{2}x - \frac{1}{4a}\sin 2ax$

177. $\int (\sin^3 ax)\, dx = -\frac{1}{3a}(\cos ax)(\sin^2 ax + 2)$

178. $\int (\sin^4 ax)\, dx = \frac{3x}{8} - \frac{\sin 2ax}{4a} + \frac{\sin 4ax}{32a}$

179. $\int (\sin^n ax)\, dx = -\frac{\sin^{n-1} ax \cos ax}{na} + \frac{n-1}{n}\int (\sin^{n-2} ax)\, dx$

180. $\int \frac{dx}{\sin^2 ax} = \int (\csc^2 ax)\, dx = -\frac{1}{a}\cot ax$

181. $\int \frac{dx}{\sin^m ax} = \int (\csc^m ax)\, dx = -\frac{1}{(m-1)a} \cdot \frac{\cos ax}{\sin^{m-1} ax} + \frac{m-2}{m-1}\int \frac{dx}{\sin^{m-2} ax}$

182. $$\int \sin(a + bx)\,dx = -\frac{1}{b}\cos(a + bx)$$

183. $$\int \frac{dx}{1 \pm \sin ax} = \mp \frac{1}{a}\tan\left(\frac{\pi}{4} \mp \frac{ax}{2}\right)$$

184. $$\int \frac{\sin ax}{1 \pm \sin ax}\,dx = \pm x + \frac{1}{a}\tan\left(\frac{\pi}{4} \mp \frac{ax}{2}\right)$$

185. $$\int \frac{dx}{(\sin ax)(1 \pm \sin ax)} = \frac{1}{a}\tan\left(\frac{\pi}{4} \mp \frac{ax}{2}\right) + \frac{1}{a}\log\tan\frac{ax}{2}$$

186. $$\int \frac{dx}{(1 + \sin ax)^2} = -\frac{1}{2a}\tan\left(\frac{\pi}{4} - \frac{ax}{2}\right) - \frac{1}{6a}\tan^3\left(\frac{\pi}{4} - \frac{ax}{2}\right)$$

187. $$\int \frac{dx}{(1 - \sin ax)^2} = \frac{1}{2a}\cot\left(\frac{\pi}{4} - \frac{ax}{2}\right) + \frac{1}{6a}\cot^3\left(\frac{\pi}{4} - \frac{ax}{2}\right)$$

188. $$\int \frac{\sin ax}{(1 + \sin ax)^2}\,dx = -\frac{1}{2a}\tan\left(\frac{\pi}{4} - \frac{ax}{2}\right) + \frac{1}{6a}\tan^3\left(\frac{\pi}{4} - \frac{ax}{2}\right)$$

189. $$\int \frac{\sin ax}{(1 - \sin ax)^2}\,dx = -\frac{1}{2a}\cot\left(\frac{\pi}{4} - \frac{ax}{2}\right) + \frac{1}{6a}\cot^3\left(\frac{\pi}{4} - \frac{ax}{2}\right)$$

190. $$\int \frac{\sin x\,dx}{a + b\sin x} = \frac{x}{b} - \frac{a}{b}\int \frac{dx}{a + b\sin x}$$

191. $$\int \frac{dx}{(\sin x)(a + b\sin x)} = \frac{1}{a}\log\tan\frac{x}{2} - \frac{b}{a}\int \frac{dx}{a + b\sin x}$$

192. $$\int \frac{dx}{(a + b\sin x)^2} = \frac{b\cos x}{(a^2 - b^2)(a + b\sin x)} + \frac{a}{a^2 - b^2}\int \frac{dx}{a + b\sin x}$$

193. $$\int \frac{\sin x\,dx}{(a + b\sin x)^2} = \frac{a\cos x}{(b^2 - a^2)(a + b\sin x)} + \frac{b}{b^2 - a^2}\int \frac{dx}{a + b\sin x}$$

194. $$\int \sqrt{1 + \sin x}\,dx = \pm 2\left(\sin\frac{x}{2} - \cos\frac{x}{2}\right),$$

[use + if $(8k - 1)\frac{\pi}{2} < x \le (8k + 3)\frac{\pi}{2}$, otherwise − ; k an integer]

195. $$\int \sqrt{1 - \sin x}\,dx = \pm 2\left(\sin\frac{x}{2} + \cos\frac{x}{2}\right),$$

[use + if $(8k - 3)\frac{\pi}{2} < x \le (8k + 1)\frac{\pi}{2}$, otherwise − ; k an integer]

196. $$\int (\cos ax)\,dx = \frac{1}{a}\sin ax$$

197. $$\int (\cos^2 ax)\,dx = \frac{1}{2a}\sin ax\cos ax + \frac{1}{2}x = \frac{1}{2}x + \frac{1}{4a}\sin 2ax$$

198. $$\int (\cos^3 ax)\,dx = \frac{1}{3a}(\sin ax)(\cos^2 ax + 2)$$

199. $$\int (\cos^4 ax)\,dx = \frac{3x}{8} + \frac{\sin 2ax}{4a} + \frac{\sin 4ax}{32a}$$

200. $$\int (\cos^n ax)\,dx = \frac{1}{na}\cos^{n-1} ax\sin ax + \frac{n-1}{n}\int (\cos^{n-2} ax)\,dx$$

201. $$\int (\cos^{2m} ax)\,dx = \frac{\sin ax}{a}\sum_{r=0}^{m-1} \frac{(2m)!(r!)^2}{2^{2m-2r}(2r+1)!(m!)^2}\cos^{2r+1} ax + \frac{(2m)!}{2^{2m}(m!)^2}x$$

202. $$\int (\cos^{2m+1} ax)\,dx = \frac{\sin ax}{a}\sum_{r=0}^{m} \frac{2^{2m-2r}(m!)^2(2r)!}{(2m+1)!(r!)^2}\cos^{2r} ax$$

203. $$\int \frac{dx}{\cos^2 ax} = \int (\sec^2 ax)\, dx = \frac{1}{a} \tan ax$$

204. $$\int \frac{dx}{\cos^n ax} = \int (\sec^n ax)\, dx = \frac{1}{(n-1)a} \cdot \frac{\sin ax}{\cos^{n-1} ax} + \frac{n-2}{n-1} \int \frac{dx}{\cos^{n-2} ax}$$

205. $$\int \cos(a + bx)\, dx = \frac{1}{b} \sin(a + bx)$$

206. $$\int \frac{dx}{1 + \cos ax} = \frac{1}{a} \tan \frac{ax}{2}$$

207. $$\int \frac{dx}{1 - \cos ax} = -\frac{1}{a} \cot \frac{ax}{2}$$

208. $$\int \frac{dx}{a + b \cos x} = \begin{cases} \dfrac{2}{\sqrt{a^2 - b^2}} \tan^{-1} \dfrac{\sqrt{a^2 - b^2} \tan \frac{x}{2}}{a + b} \\ \text{or} \\ \dfrac{1}{\sqrt{b^2 - a^2}} \log \left(\dfrac{\sqrt{b^2 - a^2} \tan \frac{x}{2} + a + b}{\sqrt{b^2 - a^2} \tan \frac{x}{2} - a - b} \right) \end{cases}$$

209. $$\int \frac{\cos ax}{1 + \cos ax}\,dx = x - \frac{1}{a}\tan\frac{ax}{2}$$

210. $$\int \frac{\cos ax}{1 - \cos ax}\,dx = -x - \frac{1}{a}\cot\frac{ax}{2}$$

211. $$\int \frac{dx}{(\cos ax)(1 + \cos ax)} = \frac{1}{a}\log\tan\left(\frac{\pi}{4} + \frac{ax}{2}\right) - \frac{1}{a}\tan\frac{ax}{2}$$

212. $$\int \frac{dx}{(\cos ax)(1 - \cos ax)} = \frac{1}{a}\log\tan\left(\frac{\pi}{4} + \frac{ax}{2}\right) - \frac{1}{a}\cot\frac{ax}{2}$$

213. $$\int \frac{dx}{(1 + \cos ax)^2} = \frac{1}{2a}\tan\frac{ax}{2} + \frac{1}{6a}\tan^3\frac{ax}{2}$$

214. $$\int \frac{dx}{(1 - \cos ax)^2} = -\frac{1}{2a}\cot\frac{ax}{2} - \frac{1}{6a}\cot^3\frac{ax}{2}$$

215. $$\int \frac{\cos ax}{(1 + \cos ax)^2}\,dx = \frac{1}{2a}\tan\frac{ax}{2} - \frac{1}{6a}\tan^3\frac{ax}{2}$$

216. $$\int \frac{\cos ax}{(1-\cos ax)^2}\,dx = \frac{1}{2a}\cot\frac{ax}{2} - \frac{1}{6a}\cot^3\frac{ax}{2}$$

217. $$\int \frac{\cos x\,dx}{a+b\cos x} = \frac{x}{b} - \frac{a}{b}\int \frac{dx}{a+b\cos x}$$

218. $$\int \frac{dx}{(\cos x)(a+b\cos x)} = \frac{1}{a}\log\tan\left(\frac{x}{2}+\frac{\pi}{4}\right) - \frac{b}{a}\int \frac{dx}{a+b\cos x}$$

219. $$\int \frac{dx}{(a+b\cos x)^2} = \frac{b\sin x}{(b^2-a^2)(a+b\cos x)} - \frac{a}{b^2-a^2}\int \frac{dx}{a+b\cos x}$$

220. $$\int \frac{\cos x}{(a+b\cos x)^2}\,dx = \frac{a\sin x}{(a^2-b^2)(a+b\cos x)} - \frac{b}{a^2-b^2}\int \frac{dx}{a+b\cos x}$$

221. $$\int \sqrt{1-\cos ax}\,dx = -\frac{2\sin ax}{a\sqrt{1-\cos ax}} = -\frac{2\sqrt{2}}{a}\cos\left(\frac{ax}{2}\right)$$

222. $$\int \sqrt{1+\cos ax}\,dx = \frac{2\sin ax}{a\sqrt{1+\cos ax}} = \frac{2\sqrt{2}}{a}\sin\left(\frac{ax}{2}\right)$$

223. $$\int \frac{dx}{\sqrt{1 - \cos x}} = \pm \sqrt{2} \log \tan \frac{x}{4},$$

[use + if $4k\pi < x < (4k + 2)\pi$, otherwise − ; k an integer]

224. $$\int \frac{dx}{\sqrt{1 + \cos x}} = \pm \sqrt{2} \log \tan \left(\frac{x + \pi}{4} \right),$$

[use + if $(4k - 1)\pi < x < (4k + 1)\pi$, otherwise − ; k an integer]

225. $$\int (\sin mx)(\sin nx)\, dx = \frac{\sin(m - n)x}{2(m - n)} - \frac{\sin(m + n)x}{2(m + n)}, \qquad (m^2 \neq n^2)$$

226. $$\int (\cos mx)(\cos nx)\, dx = \frac{\sin(m - n)x}{2(m - n)} + \frac{\sin(m + n)x}{2(m + n)}, \qquad (m^2 \neq n^2)$$

227. $$\int (\sin ax)(\cos ax)\, dx = \frac{1}{2a} \sin^2 ax$$

228. $$\int (\sin mx)(\cos nx)\, dx = -\frac{\cos(m - n)x}{2(m - n)} - \frac{\cos(m + n)x}{2(m + n)}, \qquad (m^2 \neq n^2)$$

229. $\int (\sin^2 ax)(\cos^2 ax)\, dx = -\frac{1}{32a}\sin 4ax + \frac{x}{8}$

230. $\int (\sin ax)(\cos^m ax)\, dx = -\frac{\cos^{m+1} ax}{(m+1)a}$

231. $\int (\sin^m ax)(\cos ax)\, dx = \frac{\sin^{m+1} ax}{(m+1)a}$

232. $\int \frac{\sin ax}{\cos^2 ax}\, dx = \frac{1}{a\cos ax} = \frac{\sec ax}{a}$

233 $\int \frac{\sin^2 ax}{\cos ax}\, dx = -\frac{1}{a}\sin ax + \frac{1}{a}\log\tan\left(\frac{\pi}{4} + \frac{ax}{2}\right)$

234. $\int \frac{\cos ax}{\sin^2 ax}\, dx = -\frac{1}{a\sin ax} = -\frac{\csc ax}{a}$

235. $\int \frac{dx}{(\sin ax)(\cos ax)} = \frac{1}{a}\log\tan ax$

236. $\int \frac{dx}{(\sin ax)(\cos^2 ax)} = \frac{1}{a}\left(\sec ax + \log\tan\frac{ax}{2}\right)$

237. $$\int \frac{dx}{(\sin ax)(\cos^n ax)} = \frac{1}{a(n-1)\cos^{n-1} ax} + \int \frac{dx}{(\sin ax)(\cos^{n-2} ax)}$$

238. $$\int \frac{dx}{(\sin^2 ax)(\cos ax)} = -\frac{1}{a}\csc ax + \frac{1}{a}\log\tan\left(\frac{\pi}{4} + \frac{ax}{2}\right)$$

239. $$\int \frac{dx}{(\sin^2 ax)(\cos^2 ax)} = -\frac{2}{a}\cot 2ax$$

240. $$\int \frac{\sin ax}{1 \pm \cos ax}\, dx = \mp \frac{1}{a}\log(1 \pm \cos ax)$$

241. $$\int \frac{\cos ax}{1 \pm \sin ax}\, dx = \pm \frac{1}{a}\log(1 \pm \sin ax)$$

242. $$\int \frac{dx}{(\sin ax)(1 \pm \cos ax)} = \pm \frac{1}{2a(1 \pm \cos ax)} + \frac{1}{2a}\log\tan\frac{ax}{2}$$

243. $$\int \frac{dx}{(\cos ax)(1 \pm \sin ax)} = \mp \frac{1}{2a(1 \pm \sin ax)} + \frac{1}{2a}\log\tan\left(\frac{\pi}{4} + \frac{ax}{2}\right)$$

244. $\int \frac{\sin ax}{(\cos ax)(1 \pm \cos ax)} dx = \frac{1}{a} \log (\sec ax \pm 1)$

245. $\int \frac{\cos ax}{(\sin ax)(1 \pm \sin ax)} dx = -\frac{1}{a} \log (\csc ax \pm 1)$

246. $\int \frac{\sin ax}{(\cos ax)(1 \pm \sin ax)} dx = \frac{1}{2a(1 \pm \sin ax)} \pm \frac{1}{2a} \log \tan\left(\frac{\pi}{4} + \frac{ax}{2}\right)$

247. $\int \frac{\cos ax}{(\sin ax)(1 \pm \cos ax)} dx = -\frac{1}{2a(1 \pm \cos ax)} \pm \frac{1}{2a} \log \tan \frac{ax}{2}$

248. $\int \frac{dx}{\sin ax \pm \cos ax} = \frac{1}{a\sqrt{2}} \log \tan\left(\frac{ax}{2} \pm \frac{\pi}{8}\right)$

249. $\int \frac{dx}{(\sin ax \pm \cos ax)^2} = \frac{1}{2a} \tan\left(ax \mp \frac{\pi}{4}\right)$

250. $\int \frac{dx}{1 + \cos ax \pm \sin ax} = \pm \frac{1}{a} \log\left(1 \pm \tan \frac{ax}{2}\right)$

251. $$\int \frac{dx}{a^2 \cos^2 cx - b^2 \sin^2 cx} = \frac{1}{2abc} \log \frac{b \tan cx + a}{b \tan cx - a}$$

252. $$\int \frac{\cos ax}{\sqrt{1 + b^2 \sin^2 ax}} dx = \frac{1}{ab} \log \left(b \sin ax + \sqrt{1 + b^2 \sin^2 ax}\right)$$

253. $$\int \frac{\cos ax}{\sqrt{1 - b^2 \sin^2 ax}} dx = \frac{1}{ab} \sin^{-1} (b \sin ax)$$

254. $$\int (\cos ax) \sqrt{1 + b^2 \sin^2 ax}\, dx = \frac{\sin ax}{2a} \sqrt{1 + b^2 \sin^2 ax} + \frac{1}{2ab} \log \left(b \sin ax + \sqrt{1 + b^2 \sin^2 ax}\right)$$

255. $$\int (\cos ax) \sqrt{1 - b^2 \sin^2 ax}\, dx = \frac{\sin ax}{2a} \sqrt{1 - b^2 \sin^2 ax} + \frac{1}{2ab} \sin^{-1} (b \sin ax)$$

256. $$\int (\tan ax)\, dx = -\frac{1}{a} \log \cos ax = \frac{1}{a} \log \sec ax$$

257. $\int (\cot ax)\,dx = \frac{1}{a} \log \sin ax = -\frac{1}{a} \log \csc ax$

258. $\int (\sec ax)\,dx = \frac{1}{a} \log (\sec ax + \tan ax) = \frac{1}{a} \log \tan \left(\frac{\pi}{4} + \frac{ax}{2}\right)$

259. $\int (\csc ax)\,dx = \frac{1}{a} \log (\csc ax - \cot ax) = \frac{1}{a} \log \tan \frac{ax}{2}$

260. $\int (\tan^2 ax)\,dx = \frac{1}{a} \tan ax - x$

261. $\int (\tan^3 ax)\,dx = \frac{1}{2a} \tan^2 ax + \frac{1}{a} \log \cos ax$

262. $\int (\tan^4 ax)\,dx = \frac{\tan^3 ax}{3a} - \frac{1}{a} \tan x + x$

263. $\int (\tan^n ax)\,dx = \frac{\tan^{n-1} ax}{a(n-1)} - \int (\tan^{n-2} ax)\,dx$

264. $$\int (\cot^2 ax)\,dx = -\frac{1}{a}\cot ax - x$$

265. $$\int (\cot^3 ax)\,dx = -\frac{1}{2a}\cot^2 ax - \frac{1}{a}\log \sin ax$$

266. $$\int (\cot^4 ax)\,dx = -\frac{1}{3a}\cot^3 ax + \frac{1}{a}\cot ax + x$$

267. $$\int (\cot^n ax)\,dx = -\frac{\cot^{n-1} ax}{a(n-1)} - \int (\cot^{n-2} ax)\,dx$$

Forms with Inverse Trigonometric Functions

268. $$\int (\sin^{-1} ax)\,dx = x \sin^{-1} ax + \frac{\sqrt{1 - a^2x^2}}{a}$$

269. $$\int (\cos^{-1} ax)\,dx = x \cos^{-1} ax - \frac{\sqrt{1 - a^2x^2}}{a}$$

270. $\int (\tan^{-1} ax)\, dx = x \tan^{-1} ax - \frac{1}{2a} \log (1 + a^2x^2)$

271. $\int (\cot^{-1} ax)\, dx = x \cot^{-1} ax + \frac{1}{2a} \log (1 + a^2x^2)$

272. $\int (\sec^{-1} ax)\, dx = x \sec^{-1} ax - \frac{1}{a} \log (ax + \sqrt{a^2x^2 - 1})$

273. $\int (\csc^{-1} ax)\, dx = x \csc^{-1} ax + \frac{1}{a} \log (ax + \sqrt{a^2x^2 - 1})$

274. $\int x[\sin^{-1}(ax)]\, dx = \frac{1}{4a^2}[(2a^2x^2 - 1) \sin^{-1}(ax) + ax\sqrt{1 - a^2x^2}]$

275. $\int x[\cos^{-1}(ax)]\, dx = \frac{1}{4a^2}[(2a^2x^2 - 1) \cos^{-1}(ax) - ax\sqrt{1 - a^2x^2}]$

Mixed Algebraic and Trigonometric Forms

276. $\int x(\sin ax)\,dx = \frac{1}{a^2}\sin ax - \frac{x}{a}\cos ax$

277. $\int x^2(\sin ax)\,dx = \frac{2x}{a^2}\sin ax - \frac{a^2x^2 - 2}{a^3}\cos ax$

278. $\int x^3(\sin ax)\,dx = \frac{3a^2x^2 - 6}{a^4}\sin ax - \frac{a^2x^3 - 6x}{a^3}\cos ax$

279. $\int x(\cos ax)\,dx = \frac{1}{a^2}\cos ax + \frac{x}{a}\sin ax$

280. $\int x^2(\cos ax)\,dx = \frac{2x\cos ax}{a^2} + \frac{a^2x^2 - 2}{a^3}\sin ax$

281. $\int x^3(\cos ax)\,dx = \frac{3a^2x^2 - 6}{a^4}\cos ax + \frac{a^2x^3 - 6x}{a^3}\sin ax$

282. $$\int x(\sin^2 ax)\,dx = \frac{x^2}{4} - \frac{x \sin 2ax}{4a} - \frac{\cos 2ax}{8a^2}$$

283. $$\int x^2(\sin^2 ax)\,dx = \frac{x^3}{6} - \left(\frac{x^2}{4a} - \frac{1}{8a^3}\right)\sin 2ax - \frac{x \cos 2ax}{4a^2}$$

284. $$\int x(\sin^3 ax)\,dx = \frac{x \cos 3ax}{12a} - \frac{\sin 3ax}{36a^2} - \frac{3x \cos ax}{4a} + \frac{3 \sin ax}{4a^2}$$

285. $$\int x(\cos^2 ax)\,dx = \frac{x^2}{4} + \frac{x \sin 2ax}{4a} + \frac{\cos 2ax}{8a^2}$$

286. $$\int x^2(\cos^2 ax)\,dx = \frac{x^3}{6} + \left(\frac{x^2}{4a} - \frac{1}{8a^3}\right)\sin 2ax + \frac{x \cos 2ax}{4a^2}$$

287. $$\int x(\cos^3 ax)\,dx = \frac{x \sin 3ax}{12a} + \frac{\cos 3ax}{36a^2} + \frac{3x \sin ax}{4a} + \frac{3 \cos ax}{4a^2}$$

288. $$\int \frac{\sin ax}{x^m}\,dx = -\frac{\sin ax}{(m-1)x^{m-1}} + \frac{a}{m-1}\int \frac{\cos ax}{x^{m-1}}\,dx$$

289. $$\int \frac{\cos ax}{x^m} dx = -\frac{\cos ax}{(m-1)x^{m-1}} - \frac{a}{m-1}\int \frac{\sin ax}{x^{m-1}} dx$$

290. $$\int \frac{x}{1 \pm \sin ax} dx = \mp \frac{x \cos ax}{a(1 \pm \sin ax)} + \frac{1}{a^2}\log(1 \pm \sin ax)$$

291. $$\int \frac{x}{1 + \cos ax} dx = \frac{x}{a}\tan\frac{ax}{2} + \frac{2}{a^2}\log\cos\frac{ax}{2}$$

292. $$\int \frac{x}{1 - \cos ax} dx = -\frac{x}{a}\cot\frac{ax}{2} + \frac{2}{a^2}\log\sin\frac{ax}{2}$$

293. $$\int \frac{x + \sin x}{1 + \cos x} dx = x\tan\frac{x}{2}$$

294. $$\int \frac{x - \sin x}{1 - \cos x} dx = -x\cot\frac{x}{2}$$

295. $$\int \frac{x}{\sin^2 ax} dx = \int x(\csc^2 ax)\, dx = -\frac{x \cot ax}{a} + \frac{1}{a^2}\log\sin ax$$

296. $$\int \frac{x}{\sin^n ax}\,dx = \int x(\csc^n ax)\,dx = -\frac{x \cos ax}{a(n-1)\sin^{n-1} ax} - \frac{1}{a^2(n-1)(n-2)\sin^{n-2} ax} + \frac{(n-2)}{(n-1)}\int \frac{x}{\sin^{n-2} ax}\,dx$$

297. $$\int \frac{x}{\cos^2 ax}\,dx = \int x(\sec^2 ax)\,dx = \frac{1}{a} x \tan ax + \frac{1}{a^2} \log \cos ax$$

298. $$\int \frac{x}{\cos^n ax}\,dx = \int x(\sec^n ax)\,dx = \frac{x \sin ax}{a(n-1)\cos^{n-1} ax} - \frac{1}{a^2(n-1)(n-2)\cos^{n-2} ax} + \frac{n-2}{n-1}\int \frac{x}{\cos^{n-2} ax}\,dx$$

Logarithmic Forms

299. $$\int (\log x)\,dx = x \log x - x$$

300. $$\int x(\log x)\,dx = \frac{x^2}{2} \log x - \frac{x^2}{4}$$

301. $$\int x^2(\log x)\,dx = \frac{x^3}{3}\log x - \frac{x^3}{9}$$

302. $$\int x^n(\log ax)\,dx = \frac{x^{n+1}}{n+1}\log ax - \frac{x^{n+1}}{(n+1)^2}$$

303. $$\int (\log x)^2\,dx = x(\log x)^2 - 2x\log x + 2x$$

304. $$\int \frac{(\log x)^n}{x}\,dx = \frac{1}{n+1}(\log x)^{n+1}$$

305. $$\int \frac{dx}{\log x} = \log(\log x) + \log x + \frac{(\log x)^2}{2\cdot 2!} + \frac{(\log x)^3}{3\cdot 3!} + \cdots$$

306. $$\int \frac{dx}{x\log x} = \log(\log x)$$

307. $$\int \frac{dx}{x(\log x)^n} = -\frac{1}{(n-1)(\log x)^{n-1}}$$

308. $$\int [\log(ax+b)]\,dx = \frac{ax+b}{a}\log(ax+b) - x$$

309. $\int \frac{\log(ax+b)}{x^2}\,dx = \frac{a}{b}\log x - \frac{ax+b}{bx}\log(ax+b)$

310. $\int \left[\log \frac{x+a}{x-a}\right] dx = (x+a)\log(x+a) - (x-a)\log(x-a)$

311. $\int x^n(\log X)\,dx = \frac{x^{n+1}}{n+1}\log X - \frac{2c}{n+1}\int \frac{x^{n+2}}{X}\,dx - \frac{b}{n+1}\int \frac{x^{n+1}}{X}\,dx$

where $X = a + bx + cx^2$

312. $\int [\log(x^2+a^2)]\,dx = x\log(x^2+a^2) - 2x + 2a\tan^{-1}\frac{x}{a}$

313. $\int [\log(x^2-a^2)]\,dx = x\log(x^2-a^2) - 2x + a\log\frac{x+a}{x-a}$

314. $\int x[\log(x^2 \pm a^2)]\,dx = \frac{1}{2}(x^2 \pm a^2)\log(x^2 \pm a^2) - \frac{1}{2}x^2$

315. $\int [\log (x + \sqrt{x^2 \pm a^2})]\, dx = x \log (x + \sqrt{x^2 \pm a^2}) - \sqrt{x^2 \pm a^2}$

316. $\int x[\log (x + \sqrt{x^2 \pm a^2})]\, dx = \left(\frac{x^2}{2} \pm \frac{a^2}{4}\right) \log (x + \sqrt{x^2 \pm a^2}) - \frac{x\sqrt{x^2 \pm a^2}}{4}$

317. $$\int x^m[\log (x + \sqrt{x^2 \pm a^2})]\, dx = \frac{x^{m+1}}{m+1} \log (x + \sqrt{x^2 \pm a^2}) - \frac{1}{m+1} \int \frac{x^{m+1}}{\sqrt{x^2 \pm a^2}}\, dx$$

318. $\int \frac{\log (x + \sqrt{x^2 + a^2})}{x^2}\, dx = -\frac{\log (x + \sqrt{x^2 + a^2})}{x} - \frac{1}{a} \log \frac{a + \sqrt{x^2 + a^2}}{x}$

319. $\int \frac{\log (x + \sqrt{x^2 - a^2})}{x^2}\, dx = -\frac{\log (x + \sqrt{x^2 - a^2})}{x} + \frac{1}{|a|} \sec^{-1} \frac{x}{a}$

320. $\int e^x \, dx = e^x$

321. $\int e^{-x} \, dx = -e^{-x}$

322. $\int e^{ax} \, dx = \frac{e^{ax}}{a}$

323. $\int x \, e^{ax} \, dx = \frac{e^{ax}}{a^2}(ax - 1)$

324. $\int \frac{e^{ax}}{x^m} \, dx = -\frac{1}{m-1} \frac{e^{ax}}{x^{m-1}} + \frac{a}{m-1} \int \frac{e^{ax}}{x^{m-1}} \, dx$

325. $\int e^{ax} \log x \, dx = \frac{e^{ax} \log x}{a} - \frac{1}{a} \int \frac{e^{ax}}{x} \, dx$

326. $$\int \frac{dx}{1+e^x} = x - \log(1+e^x) = \log\frac{e^x}{1+e^x}$$

327. $$\int \frac{dx}{a+be^{px}} = \frac{x}{a} - \frac{1}{ap}\log(a+be^{px})$$

328. $$\int \frac{dx}{ae^{mx}+be^{-mx}} = \frac{1}{m\sqrt{ab}}\tan^{-1}\left(e^{mx}\sqrt{\frac{a}{b}}\right), \qquad (a>0, b>0)$$

329. $$\int (a^x - a^{-x})\,dx = \frac{a^x + a^{-x}}{\log a}$$

330. $$\int \frac{e^{ax}}{b+ce^{ax}}\,dx = \frac{1}{ac}\log(b+ce^{ax})$$

331. $$\int \frac{x\,e^{ax}}{(1+ax)^2}\,dx = \frac{e^{ax}}{a^2(1+ax)}$$

332. $$\int x\,e^{-x^2}\,dx = -\tfrac{1}{2}e^{-x^2}$$

333. $$\int e^{ax}[\sin(bx)]\,dx = \frac{e^{ax}[a\sin(bx) - b\cos(bx)]}{a^2 + b^2}$$

334. $$\int e^{ax}[\sin(bx)][\sin(cx)]\,dx = \frac{e^{ax}[(b-c)\sin(b-c)x + a\cos(b-c)x]}{2[a^2 + (b-c)^2]} - \frac{e^{ax}[(b+c)\sin(b+c)x + a\cos(b+c)x]}{2[a^2 + (b+c)^2]}$$

335. $$\int e^{ax}[\cos(bx)]\,dx = \frac{e^{ax}}{a^2 + b^2}[a\cos(bx) + b\sin(bx)]$$

336. $$\int e^{ax}[\cos(bx)][\cos(cx)]\,dx = \frac{e^{ax}[(b-c)\sin(b-c)x + a\cos(b-c)x]}{2[a^2 + (b-c)^2]} + \frac{e^{ax}[(b+c)\sin(b+c)x + a\cos(b+c)x]}{2[a^2 + (b+c)^2]}$$

337. $$\int e^{ax}[\sin^n bx]\,dx = \frac{1}{a^2 + n^2b^2}\left[(a\sin bx - nb\cos bx)\,e^{ax}\sin^{n-1} bx + n(n-1)b^2 \int e^{ax}[\sin^{n-2} bx]\,dx\right]$$

338. $$\int e^{ax}[\cos^n bx]\,dx = \frac{1}{a^2 + n^2b^2}\left[(a\cos bx + nb\sin bx)\,e^{ax}\cos^{n-1} bx + n(n-1)b^2\int e^{ax}[\cos^{n-2} bx]\,dx\right]$$

339. $$\int x\,e^{ax}(\sin bx)\,dx = \frac{x\,e^{ax}}{a^2 + b^2}(a\sin bx - b\cos bx) - \frac{e^{ax}}{(a^2 + b^2)^2}[(a^2 - b^2)\sin bx - 2ab\cos bx]$$

340. $$\int x\,e^{ax}(\cos bx)\,dx = \frac{x\,e^{ax}}{a^2 + b^2}(a\cos bx + b\sin bx) - \frac{e^{ax}}{(a^2 + b^2)^2}[(a^2 - b^2)\cos bx + 2ab\sin bx]$$

Hyperbolic Forms

341. $$\int (\sinh x)\,dx = \cosh x$$

342. $\int (\cosh x)\, dx = \sinh x$

343. $\int (\tanh x)\, dx = \log \cosh x$

344. $\int (\coth x)\, dx = \log \sinh x$

345. $\int (\operatorname{sech} x)\, dx = \tan^{-1} (\sinh x)$

346. $\int \operatorname{csch} x\, dx = \log \tanh \left(\frac{x}{2}\right)$

347. $\int x(\sinh x)\, dx = x \cosh x - \sinh x$

348. $\int x^n(\sinh x)\, dx = x^n \cosh x - n \int x^{n-1}(\cosh x)\, dx$

349. $\int x(\cosh x)\,dx = x \sinh x - \cosh x$

350. $\int x^n(\cosh x)\,dx = x^n \sinh x - n \int x^{n-1}(\sinh x)\,dx$

351. $\int (\operatorname{sech} x)(\tanh x)\,dx = -\operatorname{sech} x$

352. $\int (\operatorname{csch} x)(\coth x)\,dx = -\operatorname{csch} x$

353. $\int (\sinh^2 x)\,dx = \frac{\sinh 2x}{4} - \frac{x}{2}$

354. $\int (\tanh^2 x)\,dx = x - \tanh x$

355. $\int (\tanh^n x)\,dx = -\frac{\tanh^{n-1} x}{n-1} + \int (\tanh^{n-2} x)\,dx, \qquad (n \neq 1)$

356. $\int (\operatorname{sech}^2 x)\, dx = \tanh x$

357. $\int (\cosh^2 x)\, dx = \frac{\sinh 2x}{4} + \frac{x}{2}$

358. $\int (\coth^2 x)\, dx = x - \coth x$

359. $\int (\coth^n x)\, dx = -\frac{\coth^{n-1} x}{n-1} + \int \coth^{n-2} x\, dx, \qquad (n \neq 1)$

Table of Definite Integrals

360. $$\int_1^\infty \frac{dx}{x^m} = \frac{1}{m-1}, \qquad [m > 1]$$

361. $$\int_0^\infty \frac{dx}{(1+x)x^p} = \pi \csc p\pi, \qquad [p < 1]$$

362. $$\int_0^\infty \frac{dx}{(1-x)x^p} = -\pi \cot p\pi, \qquad [p < 1]$$

363. $$\int_0^\infty \frac{x^{p-1}\,dx}{1+x} = \frac{\pi}{\sin p\pi}$$
$$= \mathrm{B}(p, 1-p) = \Gamma(p)\Gamma(1-p), \qquad [0 < p < 1]$$

364. $$\int_0^\infty \frac{x^{m-1}\,dx}{1+x^n} = \frac{\pi}{n \sin \frac{m\pi}{n}}, \qquad [0 < m < n]$$

365. $$\int_0^\infty \frac{dx}{(1+x)\sqrt{x}} = \pi$$

366. $\int_0^\infty \frac{a\,dx}{a^2 + x^2} = \frac{\pi}{2}$, if $a > 0$; 0, if $a = 0$; $-\frac{\pi}{2}$, if $a < 0$

367. $\int_0^\infty e^{-ax}\,dx = \frac{1}{a}, \qquad (a > 0)$

368. $\int_0^\infty \frac{e^{-ax} - e^{-bx}}{x}\,dx = \log\frac{b}{a}, \qquad (a, b > 0)$

369. $$\int_0^\infty x^n e^{-ax}\,dx = \begin{cases} \dfrac{\Gamma(n+1)}{a^{n+1}}, & (n > -1, a > 0) \\ \text{or} & \\ \dfrac{n!}{a^{n+1}}, & (a > 0, n \text{ positive integer}) \end{cases}$$

370. $\int_0^\infty x^n \exp(-ax^p)\,dx = \frac{\Gamma(k)}{pa^k}, \qquad \left(n > -1, p > 0, a > 0, k = \frac{n+1}{p}\right)$

371. $\int_0^\infty e^{-a^2x^2}\,dx = \frac{1}{2a}\sqrt{\pi} = \frac{1}{2a}\Gamma\left(\frac{1}{2}\right), \qquad (a > 0)$

372. $\int_0^\infty x\,e^{-x^2}\,dx = \frac{1}{2}$

373. $\int_0^\infty x^2\,e^{-x^2}\,dx = \frac{\sqrt{\pi}}{4}$

374. $\int_0^\infty x^{2n}\,e^{-ax^2}\,dx = \frac{1 \cdot 3 \cdot 5 \ldots (2n-1)}{2^{n+1}a^n}\sqrt{\frac{\pi}{a}}$

375. $\int_0^\infty x^{2n+1}\,e^{-ax^2}\,dx = \frac{n!}{2a^{n+1}}, \qquad (a > 0)$

376. $\int_0^1 x^m\,e^{-ax}\,dx = \frac{m!}{a^{m+1}}\left[1 - e^{-a}\sum_{r=0}^{m}\frac{a^r}{r!}\right]$

377. $\int_0^\infty e^{\left(-x^2 - \frac{a^2}{x^2}\right)}\,dx = \frac{e^{-2a}\sqrt{\pi}}{2}, \qquad (a \geq 0)$

378. $\int_0^\infty e^{-nx}\sqrt{x}\,dx = \frac{1}{2n}\sqrt{\frac{\pi}{n}}$

379. $\int_0^\infty \frac{e^{-nx}}{\sqrt{x}}\,dx = \sqrt{\frac{\pi}{n}}$

380. $\int_0^\infty e^{-ax}(\cos mx)\,dx = \frac{a}{a^2 + m^2}, \qquad (a > 0)$

381. $\int_0^\infty e^{-ax}(\sin mx)\,dx = \frac{m}{a^2 + m^2}, \qquad (a > 0)$

382. $\int_0^\infty x\,e^{-ax}[\sin(bx)]\,dx = \frac{2ab}{(a^2 + b^2)^2}, \qquad (a > 0)$

383. $\int_0^\infty x\,e^{-ax}[\cos(bx)]\,dx = \frac{a^2 - b^2}{(a^2 + b^2)^2}, \qquad (a > 0)$

384. $\int_0^\infty x^n e^{-ax}[\sin(bx)]\,dx = \frac{n![(a + ib)^{n+1} - (a - ib)^{n+1}]}{2i(a^2 + b^2)^{n+1}}, \qquad (i^2 = -1, a > 0)$

385. $\int_0^\infty x^n e^{-ax}[\cos(bx)]\,dx = \frac{n![(a-ib)^{n+1} + (a+ib)^{n+1}]}{2(a^2+b^2)^{n+1}}, \qquad (i^2 = -1, a > 0)$

386. $\int_0^\infty \frac{e^{-ax}\sin x}{x}\,dx = \cot^{-1} a, \qquad (a > 0)$

387. $\int_0^\infty e^{-a^2x^2}\cos bx\,dx = \frac{\sqrt{\pi}}{2a}\exp\left(-\frac{b^2}{4a^2}\right), \qquad (ab \neq 0)$

388. $\int_0^\infty e^{-t\cos\phi}\,t^{b-1}\sin(t\sin\phi)\,dt = [\Gamma(b)]\sin(b\phi), \qquad \left(b > 0, -\frac{\pi}{2} < \phi < \frac{\pi}{2}\right)$

389. $\int_0^\infty e^{-t\cos\phi}\,t^{b-1}[\cos(t\sin\phi)]\,dt = [\Gamma(b)]\cos(b\phi), \qquad \left(b > 0, -\frac{\pi}{2} < \phi < \frac{\pi}{2}\right)$

390. $\int_0^\infty t^{b-1}\cos t\,dt = [\Gamma(b)]\cos\left(\frac{b\pi}{2}\right), \qquad (0 < b < 1)$

391. $\int_0^\infty t^{b-1}(\sin t)\,dt = [\Gamma(b)]\sin\left(\frac{b\pi}{2}\right), \qquad (0 < b < 1)$

392. $\int_0^1 (\log x)^n \, dx = (-1)^n \cdot n!$

393. $\int_0^1 \left(\log \frac{1}{x}\right)^{\frac{1}{2}} dx = \frac{\sqrt{\pi}}{2}$

394. $\int_0^1 \left(\log \frac{1}{x}\right)^{-\frac{1}{2}} dx = \sqrt{\pi}$

395. $\int_0^1 \left(\log \frac{1}{x}\right)^n dx = n!$

396. $\int_0^1 x \log (1 - x) \, dx = -\frac{3}{4}$

397. $\int_0^1 x \log (1 + x) \, dx = \frac{1}{4}$

398. $\int_0^1 x^m (\log x)^n \, dx = \frac{(-1)^n n!}{(m+1)^{n+1}}, \qquad m > -1, n = 0, 1, 2, \ldots$

If $n \neq 0, 1, 2, \ldots$ replace $n!$ by $\Gamma(n + 1)$.

399. $\int_0^\infty \frac{\sin x}{x^p}\,dx = \frac{\pi}{2\Gamma(p)\sin(p\pi/2)}, \qquad 0 < p < 1$

400. $\int_0^\infty \frac{\cos x}{x^p}\,dx = \frac{\pi}{2\Gamma(p)\cos(p\pi/2)}, \qquad 0 < p < 1$

401. $\int_0^\infty \frac{1 - \cos px}{x^2}\,dx = \frac{\pi p}{2}$

402. $\int_0^\infty \frac{\sin px \cos qx}{x}\,dx = \left\{0, \quad q > p > 0; \quad \frac{\pi}{2}, \quad p > q > 0; \quad \frac{\pi}{4}, \quad p = q > 0\right\}$

403. $\int_0^\infty \frac{\cos(mx)}{x^2 + a^2}\,dx = \frac{\pi}{2|a|} e^{-|ma|}$

404. $\int_0^\infty \cos(x^2)\,dx = \int_0^\infty \sin(x^2)\,dx = \frac{1}{2}\sqrt{\frac{\pi}{2}}$

405. $\int_0^\infty \sin ax^n\,dx = \frac{1}{na^{1/n}}\Gamma(1/n)\sin\frac{\pi}{2n}, \qquad n > 1$

406. $\int_0^\infty \cos ax^n \, dx = \frac{1}{na^{1/n}} \Gamma(1/n) \cos \frac{\pi}{2n}, \qquad n > 1$

407. $\int_0^\infty \frac{\sin x}{\sqrt{x}} \, dx = \int_0^\infty \frac{\cos x}{\sqrt{x}} \, dx = \sqrt{\frac{\pi}{2}}$

408. (a) $\int_0^\infty \frac{\sin^3 x}{x} \, dx = \frac{\pi}{4}$ (b) $\int_0^\infty \frac{\sin^3 x}{x^2} \, dx \, \frac{3}{4} \log 3$

409. $\int_0^\infty \frac{\sin^3 x}{x^3} \, dx = \frac{3\pi}{8}$

410. $\int_0^\infty \frac{\sin^4 x}{x^4} \, dx = \frac{\pi}{3}$

411. $\int_0^{\pi/2} \frac{dx}{1 + a \cos x} = \frac{\cos^{-1} a}{\sqrt{1 - a^2}}, \qquad (a < 1)$

412. $\int_0^\pi \frac{dx}{a + b \cos x} = \frac{\pi}{\sqrt{a^2 - b^2}}, \qquad (a > b \geq 0)$

413. $\displaystyle\int_0^{2\pi} \frac{dx}{1 + a\cos x} = \frac{2\pi}{\sqrt{1 - a^2}}, \qquad (a^2 < 1)$

414. $\displaystyle\int_0^{\infty} \frac{\cos ax - \cos bx}{x}\, dx = \log \frac{b}{a}$

415. $\displaystyle\int_0^{\pi/2} \frac{dx}{a^2 \sin^2 x + b^2 \cos^2 x} = \frac{\pi}{2ab}$

416. $\displaystyle\int_0^{\pi/2} (\sin^n x)\, dx = \begin{cases} \displaystyle\int_0^{\pi/2} (\cos^n x)\, dx \\ \text{or} \\ \dfrac{1 \cdot 3 \cdot 5 \cdot 7 \ldots (n-1)}{2 \cdot 4 \cdot 6 \cdot 8 \ldots (n)} \dfrac{\pi}{2}, & (n \text{ an even integer, } n \neq 0) \\ \text{or} \\ \dfrac{2 \cdot 4 \cdot 6 \cdot 8 \ldots (n-1)}{1 \cdot 3 \cdot 5 \cdot 7 \ldots (n)}, & (n \text{ an odd integer, } n \neq 1) \\ \text{or} \\ \dfrac{\sqrt{\pi}}{2} \dfrac{\Gamma\left(\dfrac{n+1}{2}\right)}{\Gamma\left(\dfrac{n}{2} + 1\right)}, & (n > -1) \end{cases}$

417. $\int_0^\infty \frac{\sin mx\,dx}{x} = \frac{\pi}{2}$, if $m > 0$; 0, if $m = 0$; $-\frac{\pi}{2}$, if $m < 0$

418. $\int_0^\infty \frac{\cos x\,dx}{x} = \infty$

419. $\int_0^\infty \frac{\tan x\,dx}{x} = \frac{\pi}{2}$

420. $\int_0^\pi \sin ax \cdot \sin bx\,dx = \int_0^\pi \cos ax \cdot \cos bx\,dx = 0,\qquad (a \neq b;\ a, b \text{ integers})$

421. $\int_0^{\pi/a} [\sin(ax)][\cos(ax)]\,dx = \int_0^\pi [\sin(ax)][\cos(ax)]\,dx = 0$

422. $\int_0^\pi [\sin(ax)][\cos(bx)]\,dx = \frac{2a}{a^2 - b^2}$, if $a - b$ is odd, or 0 if $a - b$ is even

423. $\int_0^\infty \frac{\sin x \cos mx\,dx}{x}$

$$= 0, \text{ if } m < -1 \text{ or } m > 1;\ \frac{\pi}{4}, \text{ if } m = \pm 1;\ \frac{\pi}{2}, \text{ if } m^2 < 1$$

424. $\int_0^\infty \frac{\sin ax \sin bx}{x^2}\,dx = \frac{\pi a}{2}, \qquad (a \le b)$

425. $\int_0^\pi \sin^2 mx\,dx = \int_0^\pi \cos^2 mx\,dx = \frac{\pi}{2}$

426. $\int_0^\infty \frac{\sin^2 (px)}{x^2}\,dx = \frac{\pi p}{2}$

427. $\int_0^1 \frac{\log x}{1 + x}\,dx = -\frac{\pi^2}{12}$

428. $\int_0^1 \frac{\log x}{1 - x}\,dx = -\frac{\pi^2}{6}$

429. $\int_0^1 \frac{\log (1 + x)}{x}\,dx = \frac{\pi^2}{12}$

430. $\int_0^1 \frac{\log (1 - x)}{x}\,dx = -\frac{\pi^2}{6}$

431. $\int_0^1 (\log x)[\log(1+x)]\,dx = 2 - 2\log 2 - \frac{\pi^2}{12}$

432. $\int_0^1 (\log x)[\log(1-x)]\,dx = 2 - \frac{\pi^2}{6}$

433. $\int_0^1 \frac{\log x}{1-x^2}\,dx = -\frac{\pi^2}{8}$

434. $\int_0^1 \log\left(\frac{1+x}{1-x}\right)\cdot\frac{dx}{x} = \frac{\pi^2}{4}$

435. $\int_0^1 \frac{\log x\,dx}{\sqrt{1-x^2}} = -\frac{\pi}{2}\log 2$

436. $\int_0^1 x^m\left[\log\left(\frac{1}{x}\right)\right]^n dx = \frac{\Gamma(n+1)}{(m+1)^{n+1}}$, if $m+1>0, n+1>0$

437. $\int_0^1 \frac{(x^p - x^q)\,dx}{\log x} = \log\left(\frac{p+1}{q+1}\right)$, $(p+1>0, q+1>0)$

438. $\int_0^1 \frac{dx}{\sqrt{\log\left(\frac{1}{x}\right)}} = \sqrt{\pi}$

439. $\int_0^\infty \log\left(\frac{e^x+1}{e^x-1}\right) dx = \frac{\pi^2}{4}$

440. $\int_0^{\pi/2} (\log \sin x)\, dx = \int_0^{\pi/2} \log \cos x\, dx = -\frac{\pi}{2} \log 2$

441. $\int_0^{\pi/2} (\log \sec x)\, dx = \int_0^{\pi/2} \log \csc x\, dx = \frac{\pi}{2} \log 2$

442. $\int_0^{\pi} x(\log \sin x)\, dx = -\frac{\pi^2}{2} \log 2$

443. $\int_0^{\pi/2} (\sin x)(\log \sin x)\, dx = \log 2 - 1$

444. $\int_0^{\pi/2} (\log \tan x)\, dx = 0$

445. $\int_0 \log (a \pm b \cos x)\, dx = \pi \log \left(\frac{a + \sqrt{a^2 - b^2}}{2} \right), \qquad (a \geq b)$

446. $\int_0^{\pi} \log (a^2 - 2ab \cos x + b^2)\, dx = \begin{cases} 2\pi \log a, & a \geq b > 0 \\ 2\pi \log b, & b \geq a > 0 \end{cases}$

447. $\int_0^{\infty} \frac{\sin ax}{\sinh bx}\, dx = \frac{\pi}{2b} \tanh \frac{a\pi}{2b}$

448. $\int_0^{\infty} \frac{\cos ax}{\cosh bx}\, dx = \frac{\pi}{2b} \operatorname{sech} \frac{a\pi}{2b}$

449. $\int_0^{\infty} \frac{dx}{\cosh ax} = \frac{\pi}{2a}$

450. $\int_0^{\infty} \frac{x\, dx}{\sinh ax} = \frac{\pi^2}{4a^2}$

451. $\int_0^{\infty} e^{-ax} (\cosh bx)\, dx = \frac{a}{a^2 - b^2}, \qquad (0 \leq |b| < a)$

452. $\int_0^\infty e^{-ax}(\sinh bx)\,dx = \frac{b}{a^2 - b^2}, \qquad (0 \leq |b| < a)$

453. $\int_0^\infty \frac{\sinh ax}{e^{bx} + 1}\,dx = \frac{\pi}{2b}\csc\frac{a\pi}{b} - \frac{1}{2a}$

454. $\int_0^\infty \frac{\sinh ax}{e^{bx} - 1}\,dx = \frac{1}{2a} - \frac{\pi}{2b}\cot\frac{a\pi}{b}$

455. $\int_0^{\pi/2} \frac{dx}{\sqrt{1 - k^2\sin^2 x}} = \frac{\pi}{2}\left[1 + \left(\frac{1}{2}\right)^2 k^2 + \left(\frac{1\cdot 3}{2\cdot 4}\right)^2 k^4 + \left(\frac{1\cdot 3\cdot 5}{2\cdot 4\cdot 6}\right)^2 k^6 + \cdots\right], \text{ if } k^2 < 1$

456. $\int_0^{\pi/2} \sqrt{1 - k^2\sin^2 x}\,dx = \frac{\pi}{2}\left[1 - \left(\frac{1}{2}\right)^2 k^2 - \left(\frac{1\cdot 3}{2\cdot 4}\right)^2 \frac{k^4}{3} - \left(\frac{1\cdot 3\cdot 5}{2\cdot 4\cdot 6}\right)^2 \frac{k^6}{5} - \cdots\right], \text{ if } k^2 < 1$

457. $\int_0^\infty e^{-x}\log x\,dx = -\gamma = -0.5772157\ldots$

458. $\int_0^\infty e^{-x^2} \log x \, dx = -\frac{\sqrt{\pi}}{4}(\gamma + 2\log 2)$

Appendix

TABLE A.1: Areas Under the Standard Normal Curve

z	0.00	0.01	0.02	0.03	0.04	0.05	0.06	0.07	0.08	0.09
0.0	0.0000	0.0040	0.0080	0.0120	0.0160	0.0199	0.0239	0.0279	0.0319	0.0359
0.1	0.0398	0.0438	0.0478	0.0517	0.0557	0.0596	0.0636	0.0675	0.0714	0.0753
0.2	0.0793	0.0832	0.0871	0.0910	0.0948	0.0987	0.1026	0.1064	0.1103	0.1141
0.3	0.1179	0.1217	0.1255	0.1293	0.1331	0.1368	0.1406	0.1443	0.1480	0.1517
0.4	0.1554	0.1591	0.1628	0.1664	0.1700	0.1736	0.1772	0.1808	0.1844	0.1879
0.5	0.1915	0.1950	0.1985	0.2019	0.2054	0.2088	0.2123	0.2157	0.2190	0.2224
0.6	0.2257	0.2291	0.2324	0.2357	0.2389	0.2422	0.2454	0.2486	0.2517	0.2549
0.7	0.2580	0.2611	0.2642	0.2673	0.2704	0.2734	0.2764	0.2794	0.2823	0.2852
0.8	0.2881	0.2910	0.2939	0.2967	0.2995	0.3023	0.3051	0.3078	0.3106	0.3133
0.9	0.3159	0.3186	0.3212	0.3238	0.3264	0.3289	0.3315	0.3340	0.3365	0.3389
1.0	0.3413	0.3438	0.3461	0.3485	0.3508	0.3531	0.3554	0.3577	0.3599	0.3621
1.1	0.3643	0.3665	0.3686	0.3708	0.3729	0.3749	0.3770	0.3790	0.3810	0.3830
1.2	0.3849	0.3869	0.3888	0.3907	0.3925	0.3944	0.3962	0.3980	0.3997	0.4015
1.3	0.4032	0.4049	0.4066	0.4082	0.4099	0.4115	0.4131	0.4147	0.4162	0.4177
1.4	0.4192	0.4207	0.4222	0.4236	0.4251	0.4265	0.4279	0.4292	0.4306	0.4319
1.5	0.4332	0.4345	0.4357	0.4370	0.4382	0.4394	0.4406	0.4418	0.4429	0.4441
1.6	0.4452	0.4463	0.4474	0.4484	0.4495	0.4505	0.4515	0.4525	0.4535	0.4545
1.7	0.4554	0.4564	0.4573	0.4582	0.4591	0.4599	0.4608	0.4616	0.4625	0.4633
1.8	0.4641	0.4649	0.4656	0.4664	0.4671	0.4678	0.4686	0.4693	0.4699	0.4706
1.9	0.4713	0.4719	0.4726	0.4732	0.4738	0.4744	0.4750	0.4756	0.4761	0.4767
2.0	0.4772	0.4778	0.4783	0.4788	0.4793	0.4798	0.4803	0.4808	0.4812	0.4817
2.1	0.4821	0.4826	0.4830	0.4834	0.4838	0.4842	0.4846	0.4850	0.4854	0.4857
2.2	0.4861	0.4864	0.4868	0.4871	0.4875	0.4878	0.4881	0.4884	0.4887	0.4890
2.3	0.4893	0.4896	0.4898	0.4901	0.4904	0.4906	0.4909	0.4911	0.4913	0.4916
2.4	0.4918	0.4920	0.4922	0.4925	0.4927	0.4929	0.4931	0.4932	0.4934	0.4936
2.5	0.4938	0.4940	0.4941	0.4943	0.4945	0.4946	0.4948	0.4949	0.4951	0.4952
2.6	0.4953	0.4955	0.4956	0.4957	0.4959	0.4960	0.4961	0.4962	0.4963	0.4964
2.7	0.4965	0.4966	0.4967	0.4968	0.4969	0.4970	0.4971	0.4972	0.4973	0.4974
2.8	0.4974	0.4975	0.4976	0.4977	0.4977	0.4978	0.4979	0.4979	0.4980	0.4981
2.9	0.4981	0.4982	0.4982	0.4983	0.4984	0.4984	0.4985	0.4985	0.4986	0.4986
3.0	0.4987	0.4987	0.4987	0.4988	0.4988	0.4989	0.4989	0.4989	0.4990	0.4990

TABLE A.2: Poisson Distribution

Each number in this table represents the probability of obtaining at least X successes, or the area under the histogram to the right of and including the rectangle whose center is at X.

m	$X=0$	$X=1$	$X=2$	$X=3$	$X=4$	$X=5$	$X=6$	$X=7$	$X=8$	$X=9$	$X=10$	$X=11$	$X=12$	$X=13$	$X=14$
.10	1.000	.095	.005												
.20	1.000	.181	.018	.001											
.30	1.000	.259	.037	.004											
.40	1.000	.330	.062	.008	.001										
.50	1.000	.393	.090	.014	.002										
.60	1.000	.451	.122	.023	.003										
.70	1.000	.503	.156	.034	.006	.001									
.80	1.000	.551	.191	.047	.009	.001									
.90	1.000	.593	.228	.063	.013	.002									
1.00	1.000	.632	.264	.080	.019	.004	.001								
1.1	1.000	.667	.301	.100	.026	.005	.001								
1.2	1.000	.699	.337	.120	.034	.008	.002								
1.3	1.000	.727	.373	.143	.043	.011	.002								
1.4	1.000	.753	.408	.167	.054	.014	.003	.001							
1.5	1.000	.777	.442	.191	.066	.019	.004	.001							
1.6	1.000	.798	.475	.217	.079	.024	.006	.001							
1.7	1.000	.817	.507	.243	.093	.030	.008	.002							
1.8	1.000	.835	.537	.269	.109	.036	.010	.003	.001						
1.9	1.000	.850	.566	.296	.125	.044	.013	.003	.001						
2.0	1.000	.865	.594	.323	.143	.053	.017	.005	.001						
2.2	1.000	.889	.645	.377	.181	.072	.025	.007	.002						

TABLE A.2 (continued): Poisson Distribution

2.4	1.000	.909	.692	.430	.221	.096	.036	.012	.003	.001					
2.6	1.000	.926	.733	.482	.264	.123	.049	.017	.005	.001					
2.8	1.000	.939	.769	.531	.308	.152	.065	.024	.008	.002	.001				
3.0	1.000	.950	.801	.577	.353	.185	.084	.034	.012	.004	.001				
3.2	1.000	.959	.829	.620	.397	.219	.105	.045	.017	.006	.002				
3.4	1.000	.967	.853	.660	.442	.256	.129	.058	.023	.008	.003	.001			
3.6	1.000	.973	.874	.697	.485	.294	.156	.073	.031	.012	.004	.001			
3.8	1.000	.978	.893	.731	.527	.332	.184	.091	.040	.016	.006	.002			
4.0	1.000	.982	.908	.762	.567	.371	.215	.111	.051	.021	.008	.003	.001		
4.2	1.000	.985	.922	.790	.605	.410	.247	.133	.064	.028	.011	.004	.001		
4.4	1.000	.988	.934	.815	.641	.449	.280	.156	.079	.036	.015	.006	.002	.001	
4.6	1.000	.990	.944	.837	.674	.487	.314	.182	.095	.045	.020	.008	.003	.001	
4.8	1.000	.992	.952	.857	.706	.524	.349	.209	.113	.056	.025	.010	.004	.001	
5.0	1.000	.993	.960	.875	.735	.560	.384	.238	.133	0.68	.032	.014	.005	.002	.001

Reprinted from Alder, H. L. and Roessler, E. B., *Introduction to Probability and Statistics*, 6th ed., 1977. With permission of W. H. Freeman, New York.

TABLE A.3: *t*-Distribution

$-t$ 0 t

deg. freedom, f	90% ($P = 0.1$)	95% ($P = 0.05$)	99% ($P = 0.01$)
1	6.314	12.706	63.657
2	2.920	4.303	9.925
3	2.353	3.182	5.841
4	2.132	2.776	4.604
5	2.015	2.571	4.032
6	1.943	2.447	3.707
7	1.895	2.365	3.499
8	1.860	2.306	3.355
9	1.833	2.262	3.250
10	1.812	2.228	3.169
11	1.796	2.201	3.106
12	1.782	2.179	3.055
13	1.771	2.160	3.012
14	1.761	2.145	2.977
15	1.753	2.131	2.947
16	1.746	2.120	2.921
17	1.740	2.110	2.898
18	1.734	2.101	2.878
19	1.729	2.093	2.861
20	1.725	2.086	2.845
21	1.721	2.080	2.831
22	1.717	2.074	2.819
23	1.714	2.069	2.807
24	1.711	2.064	2.797
25	1.708	2.060	2.787
26	1.706	2.056	2.779
27	1.703	2.052	2.771
28	1.701	2.048	2.763
29	1.699	2.045	2.756
inf.	1.645	1.960	2.576

Reprinted from Tallarida, R. J. and Murray, R. B., *Manual of Pharmacologic Calculations with Computer Programs,* 2nd ed., 1987. With permission of Springer-Verlag, New York.

TABLE A.4: χ^2–Distribution

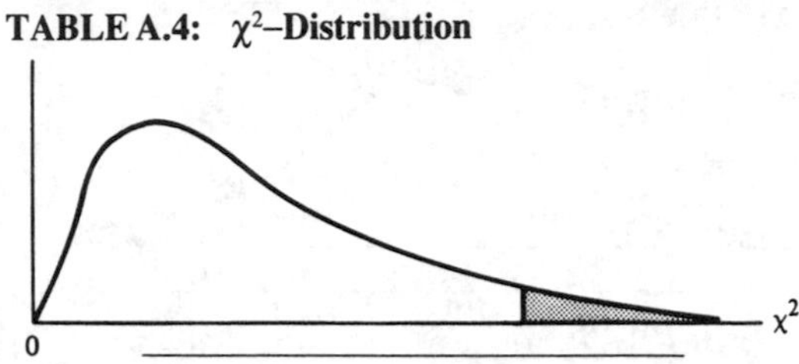

υ	0.05	0.025	0.01	0.005
1	3.841	5.024	6.635	7.879
2	5.991	7.378	9.210	10.597
3	7.815	9.348	11.345	12.838
4	9.488	11.143	13.277	14.860
5	11.070	12.832	15.086	16.750
6	12.592	14.449	16.812	18.548
7	14.067	16.013	18.475	20.278
8	15.507	17.535	20.090	21.955
9	16.919	19.023	21.666	23.589
10	18.307	20.483	23.209	25.188
11	19.675	21.920	24.725	26.757
12	21.026	23.337	26.217	28.300
13	22.362	24.736	27.688	29.819
14	23.685	26.119	29.141	31.319
15	24.996	27.488	30.578	32.801
16	26.296	28.845	32.000	34.267
17	27.587	30.191	33.409	35.718
18	28.869	31.526	34.805	37.156
19	30.144	32.852	36.191	38.582
20	31.410	34.170	37.566	39.997
21	32.671	35.479	38.932	41.401
22	33.924	36.781	40.289	42.796
23	35.172	38.076	41.638	44.181
24	36.415	39.364	42.980	45.558
25	37.652	40.646	44.314	46.928
26	38.885	41.923	45.642	48.290
27	40.113	43.194	46.963	49.645
28	41.337	44.461	48.278	50.993
29	42.557	45.722	49.588	52.336
30	43.773	46.979	50.892	53.672

Reprinted from Freund, J. E. and Williams, F.J., *Elementary Business Statistics: The Modern Approach,* 2nd ed., 1972. With permission of Prentice-Hall, Englewood Cliffs, NJ.

TABLE A.5: Variance Ratio

	F(95%)									
	n_1									
n_2	1	2	3	4	5	6	8	12	24	∞
1	161.4	199.5	215.7	224.6	230.2	234.0	238.9	243.9	249.0	254.3
2	18.51	19.00	19.16	19.25	19.30	19.33	19.37	19.41	19.45	19.50
3	10.13	9.55	9.28	9.12	9.01	8.94	8.84	8.74	8.64	8.53
4	7.71	6.94	6.59	6.39	6.26	6.16	6.04	5.91	5.77	5.63
5	6.61	5.79	5.41	5.19	5.05	4.95	4.82	4.68	4.53	4.36
6	5.99	5.14	4.76	4.53	4.39	4.28	4.15	4.00	3.84	3.67
7	5.59	4.74	4.35	4.12	3.97	3.87	3.73	3.57	3.41	3.23
8	5.32	4.46	4.07	3.84	3.69	3.58	3.44	3.28	3.12	2.93
9	5.12	4.26	3.86	3.63	3.48	3.37	3.23	3.07	2.90	2.71
10	4.96	4.10	3.71	3.48	3.33	3.22	3.07	2.91	2.74	2.54
11	4.84	3.98	3.59	3.36	3.20	3.09	2.95	2.79	2.61	2.40
12	4.75	3.88	3.49	3.26	3.11	3.00	2.85	2.69	2.50	2.30
13	4.67	3.80	3.41	3.18	3.02	2.92	2.77	2.60	2.42	2.21
14	4.60	3.74	3.34	3.11	2.96	2.85	2.70	2.53	2.35	2.13
15	4.54	3.68	3.29	3.06	2.90	2.79	2.64	2.48	2.29	2.07
16	4.49	3.63	3.24	3.01	2.85	2.74	2.59	2.42	2.24	2.01
17	4.45	3.59	3.20	2.96	2.81	2.70	2.55	2.38	2.19	1.96
18	4.41	3.55	3.16	2.93	2.77	2.66	2.51	2.34	2.15	1.92
19	4.38	3.52	3.13	2.90	2.74	2.63	2.48	2.31	2.11	1.88
20	4.35	3.49	3.10	2.87	2.71	2.60	2.45	2.28	2.08	1.84
21	4.32	3.47	3.07	2.84	2.68	2.57	2.42	2.25	2.05	1.81
22	4.30	3.44	3.05	2.82	2.66	2.55	2.40	2.23	2.03	1.78
23	4.28	3.42	3.03	2.80	2.64	2.53	2.38	2.20	2.00	1.76
24	4.26	3.40	3.01	2.78	2.62	2.51	2.36	2.18	1.98	1.73
25	4.24	3.38	2.99	2.76	2.60	2.49	2.34	2.16	1.96	1.71
26	4.22	3.37	2.98	2.74	2.59	2.47	2.32	2.15	1.95	1.69
27	4.21	3.35	2.96	2.73	2.57	2.46	2.30	2.13	1.93	1.67
28	4.20	3.34	2.95	2.71	2.56	2.44	2.29	2.12	1.91	1.65
29	4.18	3.33	2.93	2.70	2.54	2.43	2.28	2.10	1.90	1.64
30	4.17	3.32	2.92	2.69	2.53	2.42	2.27	2.09	1.89	1.62
40	4.08	3.23	2.84	2.61	2.45	2.34	2.18	2.00	1.79	1.51
60	4.00	3.15	2.76	2.52	2.37	2.25	2.10	1.92	1.70	1.39
120	3.92	3.07	2.68	2.45	2.29	2.17	2.02	1.83	1.61	1.25
∞	3.84	2.99	2.60	2.37	2.21	2.10	1.94	1.75	1.52	1.00

TABLE A.5 (continued): Variance Ratio

	F(99%)									
	n_1									
n_2	1	2	3	4	5	6	8	12	24	∞
1	4,052	4,999	5,403	5,625	5,764	5,859	5,982	6,106	6,234	6,366
2	98.50	99.00	99.17	99.25	99.30	99.33	99.37	99.42	99.46	99.50
3	34.12	30.82	29.46	28.71	28.24	27.91	27.49	27.05	26.60	26.12
4	21.20	18.00	16.69	15.98	15.52	15.21	14.80	14.37	13.93	13.46
5	16.26	13.27	12.06	11.39	10.97	10.67	10.29	9.89	9.47	9.02
6	13.74	10.92	9.78	9.15	8.75	8.47	8.10	7.72	7.31	6.88
7	12.25	9.55	8.45	7.85	7.46	7.19	6.84	6.47	6.07	5.65
8	11.26	8.65	7.59	7.01	6.63	6.37	6.03	5.67	5.28	4.86
9	10.56	8.02	6.99	6.42	6.06	5.80	5.47	5.11	4.73	4.31
10	10.04	7.56	6.55	5.99	5.64	5.39	5.06	4.71	4.33	3.91
11	9.65	7.20	6.22	5.67	5.32	5.07	4.74	4.40	4.02	3.60
12	9.33	6.93	5.95	5.41	5.06	4.82	4.50	4.16	3.78	3.36
13	9.07	6.70	5.74	5.20	4.86	4.62	4.30	3.96	3.59	3.16
14	8.86	6.51	5.56	5.03	4.69	4.46	4.14	3.80	3.43	3.00
15	8.68	6.36	5.42	4.89	4.56	4.32	4.00	3.67	3.29	2.87
16	8.53	6.23	5.29	4.77	4.44	4.20	3.89	3.55	3.18	2.75
17	8.40	6.11	5.18	4.67	4.34	4.10	3.79	3.45	3.08	2.65
18	8.28	6.01	5.09	4.58	4.25	4.01	3.71	3.37	3.00	2.57
19	8.18	5.93	5.01	4.50	4.17	3.94	3.63	3.30	2.92	2.49
20	8.10	5.85	4.94	4.43	4.10	3.87	3.56	3.23	2.86	2.42
21	8.02	5.78	4.87	4.37	4.04	3.81	3.51	3.17	2.80	2.36
22	7.94	5.72	4.82	4.31	3.99	3.76	3.45	3.12	2.75	2.31
23	7.88	5.66	4.76	4.26	3.94	3.71	3.41	3.07	2.70	2.26
24	7.82	5.61	4.72	4.22	3.90	3.67	3.36	3.03	2.66	2.21
25	7.77	5.57	4.68	4.18	3.86	3.63	3.32	2.99	2.62	2.17
26	7.72	5.53	4.64	4.14	3.82	3.59	3.29	2.96	2.58	2.13
27	7.68	5.49	4.60	4.11	3.78	3.56	3.26	2.93	2.55	2.10
28	7.64	5.45	4.57	4.07	3.75	3.53	3.23	2.90	2.52	2.06
29	7.60	5.42	4.54	4.04	3.73	3.50	3.20	2.87	2.49	2.03
30	7.56	5.39	4.51	4.02	3.70	3.47	3.17	2.84	2.47	2.01
40	7.31	5.18	4.31	3.83	3.51	3.29	2.99	2.66	2.29	1.80
60	7.08	4.98	4.13	3.65	3.34	3.12	2.82	2.50	2.12	1.60
120	6.85	4.79	3.95	3.48	3.17	2.96	2.66	2.34	1.95	1.38
∞	6.64	4.60	3.78	3.32	3.02	2.80	2.51	2.18	1.79	1.00

Reprinted from Fisher, R. A. and Yates, F., *Statistical Tables for Biological, Agricultural and Medical Research,* The Longman Group, Ltd., London, with permission.

Index